Springer Tracts in Modern Physics
Volume 141

Managing Editor: G. Höhler, Karlsruhe

Editors: J. Kühn, Karlsruhe
Th. Müller, Karlsruhe
R. D. Peccei, Los Angeles
F. Steiner, Ulm
J. Trümper, Garching
P. Wölfle, Karlsruhe

Honorary Editor: E. A. Niekisch, Jülich

Springer-Verlag Berlin Heidelberg GmbH

Springer Tracts in Modern Physics

Covering reviews with emphasis on the fields of Elementary Particle Physics, Solid-State Physics, Complex Systems, and Fundamental Astrophysics

Manuscripts for publication should be addressed to the editor mainly responsible for the field concerned:

Gerhard Höhler
Institut für Theoretische Teilchenphysik
Universität Karlsruhe
Postfach 6980
D-76128 Karlsruhe
Germany
Fax: +49 (7 21) 37 07 26
Phone: +49 (7 21) 6 08 33 75
Email: gerhard.hoehler@physik.uni-karlsruhe.de

Joachim Trümper
Max-Planck-Institut
für Extraterrestrische Physik
Postfach 1603
D-85740 Garching
Germany
Fax: +49 (89) 32 99 35 69
Phone: +49 (89) 32 99 35 59
Email: jtrumper@mpe-garching.mpg.de

Johann Kühn
Institut für Theoretische Teilchenphysik
Universität Karlsruhe
Postfach 6980
D-76128 Karlsruhe
Germany
Fax: +49 (7 21) 37 07 26
Phone: +49 (7 21) 6 08 33 72
Email: johann.kuehn@physik.uni-karlsruhe.de

Peter Wölfle
Institut für Theorie
der Kondensierten Materie
Universität Karlsruhe
Postfach 69 80
D-76128 Karlsruhe
Germany
Fax: +49 (7 21) 69 81 50
Phone: +49 (7 21) 6 08 35 90/33 67
Email: woelfle@tkm.physik.uni-karlsruhe.de

Thomas Müller
IEKP
Fakultät für Physik
Universität Karlsruhe
Postfach 6980
D-76128 Karlsruhe
Germany
Fax:+49 (7 21) 6 07 26 21
Phone: +49 (7 21) 6 08 35 24
Email: mullerth@vxcern.cern.ch

Roberto Peccei
Department of Physics
University of California, Los Angeles
405 Hilgard Avenue
Los Angeles, California 90024-1547
USA
Fax: +1 310 825 9368
Phone: +1 310 825 1042
Email: robertop@college.ucla.edu

Frank Steiner
Abteilung für Theoretische Physik
Universität Ulm
Albert-Einstein-Allee 11
D-89069 Ulm
Germany
Fax: +49 (7 31) 5 02 29 24
Phone: +49 (7 31) 5 02 29 10
Email: steiner@physik.uni-ulm.de

Werner Schweika

Disordered Alloys

Diffuse Scattering
and Monte Carlo Simulations

With 48 Figures

 Springer

Dr. Werner Schweika

Forschungszentrum Jülich
Institut für Festkörperforschung
D-52425 Jülich, Germany
Email: w.schweika@fz-juelich.de

Library of Congress Cataloging-in-Publication Data

Schweika, Werner, 1956–
Disordered alloys: diffuse scattering and Monte Carlo simulations / Werner Schweika.
p. cm. – (Springer tracts in modern physics; v. 141) Includes bibliographical references and index.

1. Electronic structure–Measurement. 2. Order-disorder in alloys–Measurement. 3. Kirkendall effect.
4. Monte Carlo method. I. Title. II. Series: Springer tracts in modern physics; 141.
QC1.S797 vol. 141 [QC176.8.E4] 539 s–dc21 97-39898 CIP

Physics and Astronomy Classification Scheme (PACS): 61.12.q, 61.12.Bt, 61.43.j,
61.43.Bn, 61.72.Ji, 64.60.Cn, 64.60.Fr, 68.35.Ct, 68.35.Md, 68.35,Rh, 81.30.Bx

ISBN 978-3-662-14763-4   ISBN 978-3-540-69550-9 (eBook)
DOI 10.1007/978-3-540-69550-9

Typesetting: Camera-ready copy by the author using a Springer T$_E$X macro-package
Cover design: *design & production* GmbH, Heidelberg
SPIN: 10575772   56/3144-5 4 3 2 1 0 – Printed on acid-free paper

für Simone, Juliane und Miriam

Preface

Structural correlations and effective interactions in disordered alloys provide a basis for the understanding of order–disorder phenomena in real alloys. Updating this classical subject, I hope that this review provides interesting and stimulating ideas and results for specialists but also for readers with more general interests. The disorder in the presented alloy systems is studied by diffuse scattering experiments and related Monte Carlo simulations, which seems to be a very fruitful combination of methods.

Emphasis is given to neutron scattering methods, though x-rays are of complementary importance. The use of neutrons offers several advantages, and the most important ones are the ability to distinguish static and dynamic properties – valuable for in-situ studies at high temperatures– and the sensitivity to scattering from light elements. The alternative description of the diffuse scattering might give more insights into the origin of local order of alloys. The experimental method of diffuse scattering provides unique and precise results on pair correlations, which challenge and hopefully inspire the theoretical progress in understanding alloy phase stability.

Structural simulations from scattering experiments are of more general interest; their limitations as well as their success rest on a surprising confinement of higher-order correlation functions by the full extent of the pair-correlation function. Hence, to give an example, the discussion of the reverse Monte Carlo method also will be of relevance for the determination of the structure of liquids. The inverse Monte Carlo method is applied to determine the most realistic effective pair interactions from the experimental data. This review discusses possible improvements and variations of this method, including, for instance, new ideas to reveal the gradient of the interaction potentials via charge fluctuations leaving their fingerprints on static displacements.

During recent years, with the availability of new experimental tools and techniques, the region near the surfaces of crystals has gained increasing interest. Near the surfaces of crystals, one may observe a large variety and complexity of ordering phenomena, different from but not independent of the bulk ordering properties. Near-surface order–disorder phenomena are viewed from results of very accurate Monte Carlo simulations, which are of importance for wetting theories, and predict the possibility of new equilibrium states of order near the surfaces, while the alloy's bulk is still in the disordered phase.

Admittedly, this review focuses on my own work and, despite some efforts, the literature review unfortunately remains incomplete. I apologize to those whose work I have overlooked.

The work presented here has benefited from various collaborations and I gratefully acknowledge M. Becker, C. Lamers, M. Pionke (Institut für Festkörperforschung, Forschungszentrum Jülich), A.E. Carlsson (Washington University, St. Louis), A. Hoser (Universität Hannover, HMI Berlin), D.P. Landau (Center of Simulational Physics, The University of Georgia, Athens, GA) and K. Binder (Johannes-Gutenberg Universität Mainz). Thanks to all colleagues of my institute for discussions, to A. Broich for technical assistance, to M. Beyss for growing single crystals. For valuable comments and reading of the manuscript it is my pleasure to acknowledge W.A. Oates (The University of Newcastle) and C. Sparks (Oak Ridge National Laboratory). My special gratitude, however, is dedicated to T. Springer (Institut für Festkörperforschung, Forschungszentrum Jülich) for his support and kind encouragement.

Jülich, July 1997   *Werner Schweika*

Contents

1. Introduction

There are many effects in real materials that may perturb an ideal crystalline order. Substitutional and interstitial defects, lattice displacements of static origin as well as those due to thermal vibrations, voids, precipitates, surfaces and grain boundaries typically exist in solids. Commercial alloys owe many of their useful properties to a mixture of such features. It is of long-standing and great fundamental interest to clarify the effects of these on the phase stability of alloys. For background information the interested reader might consult a number of excellent textbooks and conferences, e.g. [1.1–1.7]. Scattering contributions from such disorder will show up in some fashion in the diffuse scattering, i.e. the scattering between the Bragg peaks, which is shown schematically in Fig. 1.1. There will be isotropic diffuse scattering if the alloying atoms are randomly distributed on crystal lattice sites, as shown by von Laue in 1918 [1.8]. In general, the diffuse intensity will be modulated owing to nontrivial correlations among the atoms.

Two methods, diffuse scattering experiments and Monte Carlo simulations, are applied to study the local order properties and the configurational statistics in *simple* prototypes of alloy solid solutions. Diffuse scattering experiments provide a unique tool to determine quantitatively the full extent of the pair correlation functions that arise from chemical short-range order and lattice displacements among the alloying constituents. However, this diffuse scattering is typically weak and extensive, and careful experiments on single crystals are essential for such an enterprise. Most of the examples presented have been investigated by neutron scattering experiments. One reason is that this method enables us to distinguish between the elastic scattering due to the structural disorder and the inelastic scattering due to the dynamic disorder. In-situ experiments at high temperatures are often needed to study configurations in thermal equilibrium, which are of particular interest for applications of the Monte Carlo method.

There are various reasons to apply Monte Carlo simulations in this context. One advantage over analytical approaches is that this method can provide solutions that are as exact as possible, depending only on the numerical effort involved. One application is to provide a means for deducing information about structural disorder from the measured correlation functions. Furthermore, it provides a key to determine realistic interaction parameters

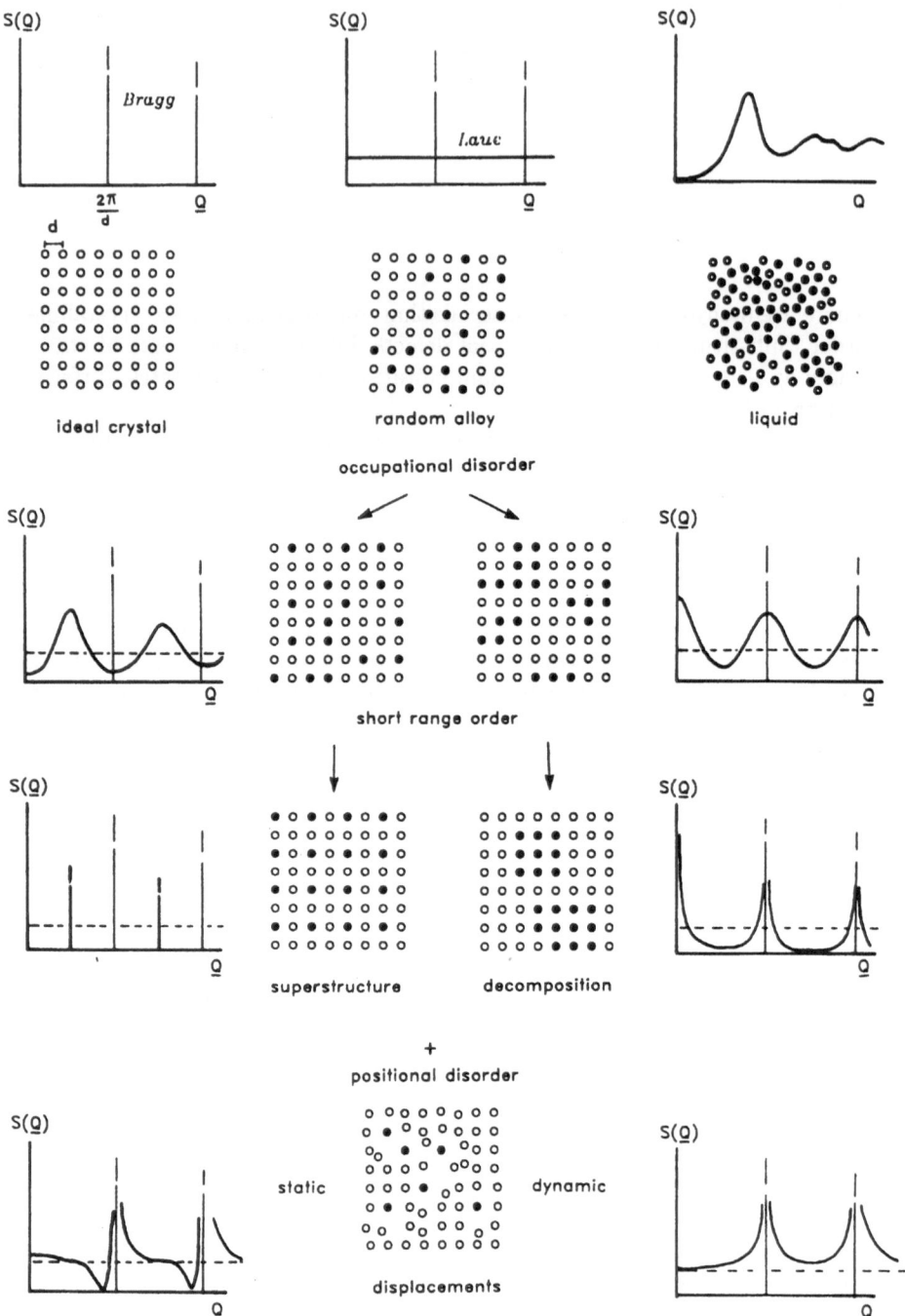

Fig. 1.1. Schematic illustration of various kinds of disorder and corresponding diffuse scattering intensities

for the alloying atoms. These parameters can be used, for instance, to calculate the phase stability with respect to coherent phase transformations. The goal of making quantitative predictions demands that we understand the often complex phenomena that drive the ordering in alloys. This implies also that we go far beyond the simple Ising Hamiltonian with only nearest-neighbor interactions which is often used to treat the statistical mechanics of alloys. Of course, the value of most such simplifying models lies in accurately analyzing universal properties and generic behavior, as will be exemplified by the studies of surface-induced ordering and disordering phenomena.

Chapter 2 reviews the background on the diffuse scattering formalism based on the kinematic approximation (see also [1.9–1.13]). In contrast to amorphous or liquid materials, the pair correlations due to chemical short-range order, size effects and displacement correlations can be separated because of the symmetry properties of the underlying mean lattice. Interestingly, the reference to the mean lattice also leads to higher-order correlation functions in the diffuse scattering cross-sections. For a thorough analysis using Kubo's cumulant expansion method we refer the reader to [1.14]. One may also draw attention to a recent review of the diffuse scattering of systems with surfaces and interfaces [1.15].

Chapter 3 discusses the Ising model approach for alloys, and the various Monte Carlo methods applied here. All these methods use the importance sampling idea used in the original Metropolis Monte Carlo method. The structural simulations on the basis of the measured pair correlation functions are performed by an algorithm which has become known from applications to liquid and amorphous systems as the "reverse Monte Carlo method". A similar method, the so-called "inverse Monte Carlo method", goes beyond the previous approach and permits one to determine interatomic effective pair interactions.

The examples presented in Chap. 4 relate closely to the author's own work and cover well-studied binary alloy systems with substitutional disorder and tendencies to ordering and clustering. Furthermore, examples are given for pseudo-binary systems with interstitial disorder, such as metal oxides and hydrides. There is an extensive literature on measurements of the local order and atomic displacements, via both x-rays and neutrons, and we refer the reader to further review articles [1.11, 1.13, 1.16–1.21]. Finally, with face-centered cubic alloy models, surface effects are studied in detail by simulations providing accurate results on possible wetting phenomena, such as surface-induced disordering. It is also found that phenomena such as surface-induced ordering may arise from the removal of frustration effects as well as the existence of effective surface fields.

2. Diffuse Scattering of Alloys

2.1 Introduction

Thermal neutron and x-ray scattering are standard techniques for the investigation of the structural properties of matter. While information about the mean structure is contained in the Bragg intensities, here we shall focus on diffuse scattering in between the Bragg peaks, which is caused by fluctuations from the mean structure due to any kind of occupational or displacive type of disorder. Before this specific topic is discussed in detail, a few general considerations are presented first.

In the usual scattering experiment, one prepares a monochromatic beam, with wave vector k_0 also defining its energy E_0, which strikes a sample. Then one measures, using a detector subtending a solid angle $\Delta\Omega$, at sufficient distances the intensity of the scattered beam, where again the wave vector k_1 (and thus E_1) needs to be determined. The kinematics of a scattering process are characterized by the conservation of momentum and energy

$$\hbar k_0 - \hbar k_1 = \hbar Q ,\tag{2.1}$$

$$E_0 - E_1 = \hbar\omega ,\tag{2.2}$$

where $\hbar Q$ is the momentum transfer and $\hbar\omega$ is the energy transfer to the sample. The scattered intensity can be expressed by a (double) differential cross-section $d^2\sigma/d\Omega d\omega(Q,\omega)$ which is the ratio of the scattered to incident particles with momentum and energy transfer, $\hbar Q$ and $\hbar\omega$ respectively, to the sample.

$W(m_0, k_0 \to m_1, k_1)$ is the transition probability (per second) in which the scattering particle goes from state $|k_0\rangle, E_0$ into $|k_1\rangle, E_1$ and the system goes from state $|m_0\rangle, E_{m_0}$ into $|m_1\rangle, E_{m_1}$ (Fermi's golden rule, [2.1])

$$
\begin{aligned}
W(m_0, k_0 \to m_1, k_1) = \frac{2\pi}{\hbar} &|\langle m_0, k_0 | H_{\text{inter}} | k_1, m_1\rangle|^2 \\
&\times \delta(E_{m_0} + E_0 - E_{m_1} - E_1) .
\end{aligned}\tag{2.3}
$$

Kinematic scattering formalism applies if the incident wave, characterized by a wave vector k and energy E, interacts only weakly with the sample (first Born approximation); this is a sufficient but not necessary requirement. All that is needed is that the measured intensity can be expressed as a scattering cross-section which is proportional to the probability of a specific scattering event.

Primary and secondary extinction, due to the scattering events of interest or other competing events respectively, and possible multiple scattering need cause only manageable (absorption) corrections. This can usually be achieved by experimental means, for instance by an appropriate choice of sample size. Since diffuse scattering is typically weak, it does not cause severe problems itself, and the kinematic scattering theory applies to the diffuse scattering. This may hold even for diffuse scattering in the case of transmission electron microscopy (TEM).[‡]

According to van Hove, the scattering cross-section is proportional to the dynamic structure factor $S(\mathbf{Q}, \omega)$, which can be interpreted as the Fourier transform of the pair correlation in space and time [2.2–2.5]:

$$\frac{d^2\sigma}{d\Omega d\omega}(\mathbf{Q}, \omega) = \frac{k_1}{k_0}(2\pi)^{-1} \int \langle \frac{1}{N} \sum_{m,n} b_m b_n e^{-i\mathbf{Q}\cdot\mathbf{r}_m(0)} e^{i\mathbf{Q}\cdot\mathbf{r}_n(t)} e^{-i\omega t} \rangle \, dt \,. \quad (2.4)$$

The wavelength of thermal and cold neutrons is large compared with the size of the nuclei, with which they strongly interact. Hence the neutron scattering amplitudes or lengths b_m are constants. The sum is over pairs of scattering centers m, n whose phase difference is determined from the projection of the difference of paths $\mathbf{r}_l = \mathbf{r}_n - \mathbf{r}_m$ onto the scattering vector $\mathbf{Q}$. Introducing a pair-correlation function $G(\mathbf{r}, t)$ permits one to take the scattering amplitudes b out of the average:

$$\frac{d^2\sigma}{d\Omega d\omega}(\mathbf{Q}, \omega) = b^2 \frac{k_1}{k_0}(2\pi)^{-1} \int\int G(\mathbf{r}, t)\, e^{i\mathbf{Q}\cdot\mathbf{r}} e^{-i\omega t} d\mathbf{r} dt \quad (2.5)$$

which separates the cross-section into a part which is dependent on the actual probe, $b^2 k_1/k_0$ for neutrons, and a part which depends only on the properties of the scattering system. This second part is the Fourier transform of the spatial and time-dependent pair-correlation function:

$$\frac{d^2\sigma}{d\Omega d\omega}(\mathbf{Q}, \omega) = b^2 \frac{k_1}{k_0}(2\pi)^{-1} \int S(\mathbf{Q}, t)\, e^{-i\omega t} dt$$

$$= b^2 \frac{k_1}{k_0}(2\pi)^{-1} S(\mathbf{Q}, \omega)\,. \quad (2.6)$$

There is a simple classical interpretation of $G(\mathbf{r}, t)$: It is the probability that there is a particle at spatial position $\mathbf{r}$ and time t and a particle (the same or a different one) at the chosen origin in space and time.

It is straightforward to generalize to the case of different kinds of atoms, and one obtains:

[‡] Quantitative diffuse scattering studies as presented here for neutrons may become feasible for TEM also because of recent instrumental developments (in particular, energy filters to discriminate plasmon excitations).

$$\frac{\mathrm{d}^2\sigma}{\mathrm{d}\Omega\mathrm{d}\omega}(\boldsymbol{Q},\omega) = \sum_{i,j} b^i b^j \frac{k_1}{k_0} (2\pi)^{-1} \int \int G_{ij}(\boldsymbol{r},t)\, \mathrm{e}^{\mathrm{i}\boldsymbol{Q}\cdot\boldsymbol{r}} \mathrm{e}^{-\mathrm{i}\omega t} \mathrm{d}\boldsymbol{r}\mathrm{d}t$$

$$= \sum_{i,j} b^i b^j \frac{k_1}{k_0} (2\pi)^{-1} S_{ij}(\boldsymbol{Q},\omega)\,. \tag{2.7}$$

It should be emphasized here, and will be further discussed below, that within the first Born approximation, the scattering cross-section reveals neither more nor less than the (Fourier transforms) of pair-correlations between the scattering particles of the sample system.

From now on we shall confine ourselves to the elastic scattering only, since we are primarily interested here in time-independent structural properties, i.e., pair correlations $G(\boldsymbol{r}, t \to \infty)$, which give the probabilities that there are atoms at distances $\boldsymbol{r}$ independent of time, i.e. for

$$|\boldsymbol{k}_0| = |\boldsymbol{k}_1|, \quad \hbar\omega = 0\,. \tag{2.8}$$

Within a given experimental energy resolution one integrates over a finite "window", which is typically quite different and much better for neutrons compared with x-rays. With neutrons and a resolution of say 1 meV, it is easy to separate most of the inelastic scattering due to phonons (and also magnons), while this is not the case when using x-rays. For x-rays one only can readily achieve a sufficient resolution to discriminate electronic excitations and most of the Compton scattering. Therefore, thermal vibrations will usually contribute to the diffuse scattering of x-rays and will be seen as in a "snapshot" by the static structure factor

$$S(\boldsymbol{Q}) = (2\pi)^{-1} \int_{-\infty}^{\infty} \mathrm{d}\omega\, S(\boldsymbol{Q},\omega)|_{\boldsymbol{Q}}\,. \tag{2.9}$$

In particular, if one considers a liquid, where all the atoms are moving, this static structure factor is of interest since it yields, upon back Fourier transformation, the instantaneous pair-correlation function.

However, the diffuse scattering of a solid is inherently not truly elastic but quasi-elastically broadened, as long as there is any diffusion. Thus with the resolution mentioned one measures the integrated quasi-elastic diffuse scattering with neutrons. If one is puristic, there is only elastic scattering in the form of a δ-function satisfying the Bragg condition, where $\boldsymbol{Q} = \boldsymbol{G}$, a reciprocal lattice vector of an infinite crystal. Since here the scattering of neutrons is of particular interest, the diffuse scattering cross-section $(\mathrm{d}\sigma/\mathrm{d}\Omega)(\boldsymbol{Q})_{\mathrm{diffuse}}$ discussed below will refer to the specific case of the integrated cross-section over the narrow quasi-elastic distribution. Hence this cross-section does not resolve the slow diffusive atomic movements. In cases where the differences from the true elastic scattering or the static structure factor, as seen by x-rays or electrons, are of interest, these will be noted explicitly.

In the case of x-ray scattering one has to replace the neutron scattering lengths b by the x-ray atomic form factors $f(Q)$ and to account for the polarizations $\boldsymbol{e}_i$ and $\boldsymbol{e}_f$ of the initial and final photons. The Thomson cross-section of a free electron is

$$\frac{d\sigma}{d\Omega}\bigg|_{\text{Thomson}} = r_e^2 (e_i \cdot e_f)^2 , \tag{2.10}$$

where the classical electron radius $r_e = 0.282 \times 10^{-12}$ cm can be regarded as the electron's scattering factor. The scattering factor of an atom is given by the Fourier transform of the electron density distribution $\rho(r)$,

$$f_o(Q) = \frac{r_e}{e} \int_V \rho(r) e^{iQ \cdot r} dr , \tag{2.11}$$

where at $Q = 0$, f/r_e equals the atomic number Z, and for larger Q, $f_o(Q)$ decreases and is typically assumed to be isotropic for studies of disordered materials. Close to the absorption edges, the scattering factor is influenced by resonant transitions to unfilled bound states:

$$f(Q, \omega) = f_o(Q) + f'(\omega) + i f''(\omega) . \tag{2.12}$$

Such resonant changes of the scattering amplitudes offer an opportunity to enhance or to vary the scattering contrast for investigations of disordered alloys [2.6]. The change of the real part of f' can be measured by interferometry, although more typically one would measure f'' via the absorption properties and determine f' using the Kramers–Kronig relation [2.7].

2.2 Disordered Solid Alloys

Now we consider a solid disordered alloy which has a mean lattice and which may contain, in general, many different elements. It is convenient to introduce a site occupation operator c_m^i which equals one if there is an atom of kind i at site r_m and which is zero otherwise. Here, it is implicitly assumed that the jump time of atoms is negligible compared with the time of residence. The average concentrations are denoted by $c^i = \langle c_m^i \rangle$.

The positional type of disorder is described by the displacements u_m with respect to the positions of the mean lattice R_m; $r_m = R_m + u_m$. There are static displacements $u(t \to \infty)$ due to the different sizes and interactions between the atoms and in addition, the time-dependent displacements $u(t)$ due to thermal vibrations (phonons).

The total scattering amplitude is

$$A(Q) = \sum_i \sum_m b^i c_m^i e^{-iQ \cdot r_m} . \tag{2.13}$$

The configurational average of the total scattering amplitude gives

$$\langle A(Q) \rangle = \sum_m e^{-iQ \cdot R_m} \sum_i b^i \langle c^i e^{-iQ \cdot u_m} \rangle , \tag{2.14}$$

where one can define the Debye–Waller factor (in the harmonic approximation) as

$$e^{-W^i(Q)} \equiv \langle e^{-iQ \cdot u_m^i} \rangle \quad (\approx e^{-\frac{1}{2}\langle (Q \cdot u_m^i)^2 \rangle}) , \tag{2.15}$$

which includes contributions from both the static and the dynamic displacements. For simplicity, it is assumed that the dynamic displacements due to phonons behave independently from the static displacements, and that the occupation and displacement variables for the same sites are statistically uncorrelated. The Debye–Waller factor can be regarded as the result of the self-correlation function, and thus it is similar to the atomic form factor for the x-ray scattering amplitude, which is Q-dependent because of the spatial distribution of the electron density of the scattering atom.

Hence, to simplify the notation, we include this form factor in the atomic scattering amplitudes, as was implicitly already done in (2.14),

$$b^i e^{-W^i(Q)} \rightarrow b^i . \tag{2.16}$$

One may note that the Debye–Waller factor describes physically the loss of constructive interference due to displacements of static or dynamic origin. For the part of the energy-integrated intensity (as seen in the static structure factor) which results from the self-correlation function, all such contributions cancel identically [2.3–2.5, 2.8]. In particular, for the elastic part of this intensity, the contributions from the static displacements will cancel.

The Bragg scattering (per atom) is determined by the properties of the mean lattice,

$$\frac{d\sigma}{d\Omega}(Q)_{\text{Bragg}} = \frac{1}{N} |\langle A(Q) \rangle|^2 , \tag{2.17}$$

while the diffuse scattering (per atom) results from the fluctuating part of the total scattering amplitude,

$$\frac{d\sigma}{d\Omega}(Q)_{\text{diffuse}} = \frac{1}{N} \langle |A(Q) - \langle A(Q) \rangle|^2 \rangle$$
$$= \frac{1}{N} (\langle |A(Q)|^2 \rangle - |\langle A(Q) \rangle|^2) . \tag{2.18}$$

Furthermore, in order to expand the phase factors we shall assume that the displacements are sufficiently small, more precisely that $Q \cdot u_m \ll 1$.

Following the usual harmonic approximation for phonon scattering [2.3–2.5], the static and dynamic displacement terms in $\langle |A(Q)|^2 \rangle$ may be expanded as

$$\langle c_m^i c_n^j e^{iQ \cdot (u_n^i - u_m^j)} \rangle = \langle c_m^i c_n^j (1 + iQ \cdot u_n^i - \frac{1}{2}(Q \cdot u_n^i)^2 ...)$$
$$\times (1 - iQ \cdot u_m^j - \frac{1}{2}(Q \cdot u_m^j)^2 ...) \rangle$$

$$\approx e^{-W^i} e^{-W^j} \langle c_m^i c_n^j (1 + iQ \cdot (u_n^i - u_m^j) + (Q \cdot u_m^i)(Q \cdot u_n^j) ...) \rangle . \tag{2.19}$$

The linear term in the displacements arises only from the static displacements.

Assuming a homogeneous system, we introduce the difference vector $R_l = R_n - R_m$ and relative displacements $u_l = u_n - u_m$. Then, from (2.18) and (2.19) we obtain

$$\frac{d\sigma}{d\Omega}(Q)_{\text{diffuse}} = \sum_{i,j} b^i b^j \sum_l e^{iQ \cdot R_l} \langle c_m^i c_{m+l}^j - c^i c^j$$

$$+ c_m^i c_{m+l}^j (iQ \cdot u_l + (Q \cdot u_m)(Q \cdot u_{m+l}) + ...] \rangle , \quad (2.20)$$

where again the Debye–Waller factors have been included in the scattering amplitudes.[‡] The terms explicitly written in (2.20) result from a linear expansion of the phase factors of the amplitudes. These terms are consistent with the harmonic approximation, which, however, is less justified for the static displacements than it is for the usual treatment of phonons as pointed out by Krivoglaz [2.9]. The static displacements in concentrated alloys are not likely to follow a unimodal Gaussian distribution. Therefore, Dietrich and Fenzl [1.14] discussed in detail the use of the cumulant expansion (Kubo [2.10]), instead of the usual Taylor expansion, because of its superior convergence property. For a further discussion of the Debye–Waller factors we also recommend of Dietrich and Fenzl [1.14] and Schönfeld [1.18].

At first sight, this formulation – where, instead of correlations between the *particles*, we have introduced correlation functions in terms of the *occupations* of and the *displacements* from the mean lattice sites – does not seem to yield a simpler description compared to the scattering of liquids. For the latter case and a binary system, for instance, from (2.7) one obtains simply three (static) partial structure factors and (instantaneous) pair-correlation functions for the *particles* (as in the typical formulation of Faber and Ziman [2.11] for liquids and amorphous materials) $S_{ij}(Q)$ and $G_{ij}(r)$ respectively, where r is a continuous variable. However, it has to be emphasized that, as (2.20) shows, the symmetry properties of a crystal separate the scattering terms with respect to the powers of the displacements involved. As discussed further below, the first term describes the chemical short-range order, the second gives approximately the size-effect scattering and the term quadratic in u corresponds to correlations in density fluctuations. Equation (2.20) already has much apparent similarity to the scattering formulation of Bathia and Thornton for liquids [2.12]. This equivalent description expresses the scattering in terms of correlations of chemical and density fluctuations. Just as for the liquid case, and as will be shown for the present case, the occupational (chemical) and positional (density) fluctuations should be associated with the difference of the scattering amplitudes $\Delta b = b_i - b_j$ and the

---

[‡] The term linear in $Q \cdot u$ in (2.20) is a configurational average and agrees with the two terms in (3.13-15) of [1.14] which are both cumulant averages. Here, these contributions related to the two- and three-point correlations will be distinguished in Sect. 2.4. Furthermore, the usual static Debye–Waller factor was introduced by assuming that the occupation and displacement variables for the same sites are statistically uncorrelated.

mean scattering amplitudes $\langle b \rangle$ respectively. The chemical short-range order terms should become "visible" by the square of the scattering contrast $|\Delta b|^2$, the size-effect scattering by $\Delta b^* \langle b \rangle$, and the displacement–displacement correlations by the weight $\langle b \rangle^2$. There will also be differences which result from the fact that the definition of the correlation functions with respect to the mean lattice has introduced higher-order correlations in (2.20). According to Dietrich and Fenzl [1.14] the diffuse scattering of disordered crystals also contains, via the displacement terms, information on higher-order *particle* correlations, although at first sight this seems to contradict the first Born approximation and needs to be discussed further below.

Beside this point of principal interest, it will be shown that, unlike the case of liquids, for solid binary alloys the true pair correlations can be separated because of the lattice symmetry properties. Therefore, it is not necessary to measure independently the different partial scattering cross-sections by varying the scattering contrast. This latter could be done using either different isotopes for neutron scattering experiments, anomalous x-ray synchrotron scattering or combined experiments with x-rays and neutrons.

2.3 Short-Range Order

In the absence of any displacements,(2.20) would reduce to the contribution of the occupational fluctuations, i.e. the chemical short-range order among the species. It is useful to note a few properties. This scattering contribution

$$\frac{d\sigma}{d\Omega}(Q)_{\text{SRO}} = \sum_{i,j} b^i b^j \sum_l g^{ij}(R_l) \, e^{iQ \cdot R_l} \qquad (2.21)$$

is given by the Fourier transform of the correlation function for the occupations of two sites with respect to the mean occupations,

$$g^{ij}(R_l) \equiv \langle \Delta c_m^i \Delta c_{m+l}^j \rangle \qquad (2.22)$$

$$= \langle c_m^i c_{m+l}^j \rangle - c^i c^j \ , \ \text{with } \Delta c_m^i \equiv c_m^i - c^i \ . \qquad (2.23)$$

In particular, the self-correlation function $g^{ij}(0)$ for $R_l = 0$ is

$$g^{ij}(0) = c^i \delta_{ij} - c^i c^j \ . \qquad (2.24)$$

We note the properties

$$\sum_j g^{ij}(R_l) = 0 \qquad (2.25)$$

and

$$g^{ii}(R_l) = \sum_{j \neq i} -g^{ij}(R_l) \ . \qquad (2.26)$$

One may define a short-range order parameter $\alpha^{ij}(R_l)$ which is the two-particle correlator normalized by the self-correlation function

$$\alpha^{ij}(R_l) \equiv \frac{g^{ij}(R_l)}{g^{ij}(0)}$$

$$= 1 - \frac{\langle c_m^i c_{m+l}^j \rangle}{c^i c^j} . \tag{2.27}$$

Because of (2.25), $\alpha^{ij}(R_l)$ does not depend on the pairing of species ij for a binary system and is commonly called the Warren–Cowley short-range order parameter $\alpha(R_l)$:

$$\alpha^{(ij)}(R_l) = \frac{g^{ij}(R_l)}{g^{ij}(0)} = \frac{g^{ii}(R_l)}{g^{ii}(0)} . \tag{2.28}$$

Using (2.26) the scattering due to short-range order (2.21) can be rewritten as

$$\frac{d\sigma}{d\Omega}(Q)_{\text{SRO}} = \sum_{\substack{i,j \\ i>j}} |b^i - b^j|^2 \sum_l g^{ij}(R_l) \, e^{iQ \cdot R_l}$$

$$= \sum_{\substack{i,j \\ i>j}} |b^i - b^j|^2 \sum_l c^i c^j \sum_l \alpha^{ij}(R_l) \, e^{iQ \cdot R_l}$$

$$= \sum_{\substack{i,j \\ i>j}} |b^i - b^j|^2 c^i c^j \sum_l \alpha^{ij}(q)$$

$$= \sum_{\substack{i,j \\ i>j}} |b^i - b^j|^2 N \langle \Delta c_i(q) \Delta c_j^*(q) \rangle . \tag{2.29}$$

Because of the translational symmetry of the mean crystal, $\alpha^{ij}(Q) = \alpha^{ij}(q)$ is a periodic function in reciprocal space, using $q = \text{Mod}(Q, G)$. Since $\alpha^{ij}(R_l)$ is always real, the Fourier transform $\alpha^{ij}(q) = \alpha^{ij}(-q)$ is a symmetric (even) function. Furthermore, we note that the scattering contribution due to chemical short-range order can be seen with the weight $|b^i - b^j|^2$, i.e. the square of the scattering contrast.

There is a single independent function describing the chemical short-range order for each type of distinct pairing $\{ij\}$, which holds also for liquids or other non-crystalline matter.

Generally, $\alpha^{ij}(R_l)$ may vary between -1 and $+1$, unless there are further geometrical restrictions. A *negative sign corresponds to an ordering tendency* with unlike neighboring atoms and a *positive sign to a clustering tendency* of like neighbors. If one considers a typical solid solution in the disordered phase, there is usually some degree of short-range order to be seen, depending on the effective interaction between the species because of their different chemical properties. In case of short-range interactions, the asymptotic decay of the short-range order $\alpha^{ij}(R_l)$ is of exponential form (Ornstein–Zernike).

In the case in which the scattering atoms are randomly distributed on the lattice sites, all the distinct pair correlations vanish, and only the self-correlation ($\alpha(0) = 1$) contributes to the diffuse scattering. This is the so-called Laue scattering

$$\frac{d\sigma}{d\Omega}_{\text{Laue}} = \sum_{\substack{i,j \\ i>j}} |b^i - b^j|^2 c^i c^j. \tag{2.30}$$

The incoherent neutron cross-section for isotopic mixtures may serve as a good example. Note that the scattering of neutrons depends on the properties of the nuclei and thus varies, and indeed often strongly, between the isotopes of one element.

Apart from the atomic scattering factors this (quasi-elastic) scattering is isotropic. Recall that the static and dynamic Debye–Waller factors have been included in the scattering amplitudes. In addition, for x-rays one has to account for atomic form factors. However, there is a sum rule for those elastic and inelastic contributions which merely stem from the autocorrelation function. Therefore, in the case of energy integration as in typical x-ray experiments, no Debye–Waller factor applies to the Laue intensity and only scattering due to distinct pair correlations is affected. Similarly, there should be no static Debye–Waller factor for the Laue scattering, since it has to cancel with all contributions from the even-order displacement terms to the Laue scattering. Because the static displacements contribute to the scattering independently of the energy resolution, it holds for both x-ray diffraction and (quasi-)elastic neutron scattering.

However, strictly speaking all the terms need to have Debye–Waller factors. It may merely be convenient, in view of cancellations with second-order displacement scattering contributions (see below), not to apply a Debye–Waller correction and to ignore consistently these latter terms.

There are no difficulties in the generalization of the formulation above to the case of disorder in which different sublattices (s, s') are involved. Of course, a normalization of the inter-sublattice correlations by the self-correlation function is not possible because it is not defined. One choice for obtaining suitable short-range order parameters would be to normalize with respect to the trace of $g_{ss'}^{ij}(\boldsymbol{R}_l)$:

$$\bar{\alpha}_{ss'}^{ij}(\boldsymbol{R}_l) = \frac{\sum_{s,s'} g_{ss'}^{ij}(\boldsymbol{R}_l)}{\sum_s g_{ss}^{ij}(0)} \tag{2.31}$$

(e.g. [2.13, 4.46]). This yields the parameters that can be obtained directly as Fourier coefficients from the analysis of the diffuse scattering data, and means the summing over all possible different contributions from the sublattices involved to the same distance $\boldsymbol{R}_l$. Again $\bar{\alpha}_{ss'}^{ij}(0) \equiv 1$, and this definition also implies further restrictions on $\bar{\alpha}_{ss'}^{ij}(\boldsymbol{R}_l)$ in addition to the maximum interval $[-1, 1]$. Alternatively [2.15], and in particular where it is possible to separate the single correlation functions $g_{ss'}^{ij}(\boldsymbol{R}_l)$ by symmetry properties, one may define the parameters $\alpha_{ss'}^{ij}(\boldsymbol{R}_l)$ as

$$\alpha_{ss'}^{ij}(\boldsymbol{R}_l) = \frac{g_{ss'}^{ij}(\boldsymbol{R}_l)}{\sqrt{g_{ss}^{ii}(0)g_{s's'}^{jj}(0)}} \tag{2.32}$$

which conserves the property $-1 \leq \alpha \leq 1$ of the Cowley–Warren short-range order parameters, and which is compatible with the common normalization of a correlation matrix.

The symmetry properties of the intra- and inter-sublattice correlations may be different or may be not, and therefore, eventually, the Fourier transforms $\alpha^{ij}_{ss'}(q)$ can be distinguished. However, the situation in which one can distinguish all physically different short-range order correlations is more an exception than the usual case. For example, and as will be discussed in Chap. 4 in the case of the pseudo-binary system $Fe_{1-x}O$ one has to consider not only the cation-vacancy disorder on the fcc cation sublattice but also on the sc tetrahedral interstitial sublattice having a common subset of difference vectors R_l. At least for the $Fe_{1-x}O$ case, one can favorably distinguish the dominating correlations within the fcc cation sublattice from the inter-sublattice correlations having reduced bcc symmetry properties. (Of course, it would not help to vary the scattering amplitudes.) That *scattering experiments cannot in principle reveal the complete short-range order in all cases* may be demonstrated by another example. Considering the system $YBa_2Cu_3O_{6+x}$, here it will be impossible to distinguish the two types of oxygen-vacancy correlations which either have a Cu atom in between or not.

2.4 Lattice Displacements

The next terms in (2.20) contain the displacement variable u to increasing order. The term linear in $Q \cdot u$ is commonly, but also loosely, called the "size-effect" scattering term. This should refer to a two-point correlation in which an occupational fluctuation is related to a positional fluctuation and vice versa. In fact, the configurational average is to be taken over products of the occupations of the two considered sites *and* terms involving the displacements. In this sense, all these correlations are not pure pair correlations any more, but correlations of higher order, which is a consequence of the choice of variables c^i_m and u referring to the deviations from the mean lattice:

$$\frac{d\sigma}{d\Omega}(Q)_{\substack{\text{lin.} \\ \text{disp.}}} = \sum_{i,j} b^i b^j \sum_l \langle c^i_m c^j_{m+l} iQ \cdot u_l \rangle e^{iQ \cdot R_l} . \tag{2.33}$$

Summarizing the linear combination of the displacement terms, one may define a function $\gamma(R_l)$, with

$$\gamma(R_l) \equiv (\frac{d\sigma}{d\Omega}_{\text{Laue}})^{-1} \sum_{i,j} b^i b^j \langle c^i_m c^j_{m+l} \cdot u_l \rangle, \tag{2.34}$$

and rewrite (2.33) as

$$\frac{d\sigma}{d\Omega}(Q)_{\substack{\text{lin.} \\ \text{disp.}}} = \frac{d\sigma}{d\Omega}_{\text{Laue}} \sum_l iQ \cdot \gamma(R_l) e^{iQ \cdot R_l}$$

$$= \frac{d\sigma}{d\Omega}_{\text{Laue}} iQ \cdot \gamma(q) \,. \tag{2.35}$$

Of course the intensity has to be real, and therefore $\gamma(q) = -\gamma(-q)$ is an imaginary antisymmetric function which reflects the fact that the static displacements have inversion symmetry properties if this is true for the underlying lattice. One may also think of static displacements which do not have the inversion symmetry of the lattice and thus do not contribute to the linear displacement scattering. These are phonon-like displacement patterns as, for instance, described by the cooperative Jahn–Teller effect. For example, a zig-zag-like static displacement pattern decorated periodically with A and B atoms can be regarded as a *frozen* optical phonon of transverse polarization, and hence will contribute to the displacement scattering of second order (u^2).

For convenience of notation, one may denote the displacements with respect to the occupations and define, for instance, average displacements $\langle u_l^{ij} \rangle$ for sites separated by R_l:

$$\langle u_l^{ij} \rangle = \langle c_m^i c_{m+l}^j (u_{m+l}^j - u_m^i) \rangle / \langle c_m^i c_{m+l}^j \rangle \,, \tag{2.36}$$

with $\langle u_l^{ij} \rangle = \langle u_l^{ji} \rangle$ and $\langle u_0^{ij} \rangle = 0$.

Furthermore, because of the existence of a mean lattice the static displacements have the properties that $\sum_m u_m^i = 0$ and

$$\sum_{i,j} \langle c_m^i c_{m+l}^j \rangle \langle u_l^{ij} \rangle = 0 \,. \tag{2.37}$$

Using (2.37), for a binary solid solution of A and B atoms, (2.33) can be rewritten as

$$\frac{d\sigma}{d\Omega}(Q)_{\substack{\text{lin.} \\ \text{disp.}}} = (b^A - b^B) \sum_l e^{iQ \cdot R_l} iQ \cdot (b^A \langle c_m^A c_{m+l}^A \rangle \langle u_l^{AA} \rangle$$
$$- b^B \langle c_m^B c_{m+l}^B \rangle \langle u_l^{BB} \rangle) \,. \tag{2.38}$$

As can be seen, without any scattering contrast $\Delta b = b^i - b^j$, the whole scattering term in (2.38), which is linear in the displacements, would have to vanish. Because of (2.37) there are two independent linear functions in the displacement scattering term of (2.20), different from the single independent function for the short-range order scattering term. Considering a multi-component alloy, there will be $(\sum_i i) - 1$ linearly independent displacement functions, which are vectors of usually three independent components and, for directions of higher symmetry, this reduces to two or one independent component. However, it already takes considerable experimental effort to distinguish the two terms for a binary alloy. In order to separate these, one needs to vary the scattering contrast. This can be done, for instance, by (i) combining neutron and x-ray data, (ii) using different isotopes for neutron scattering, (iii) making use of a sufficiently different form factor variation with Q for the different species in the case of x-rays, and (iv) using synchrotron radiation far and near the absorption edges, which also causes variations in the real parts of the atomic x-ray scattering amplitudes.

It is straightforward to introduce different sublattices, as has been done for the short-range order scattering. The displacement terms for inter- and intra-sublattice correlations can be distinguished if they have different symmetry properties.

As seen above, there is no linear displacement scattering from one sublattice occupied by atoms which show only a negligible scattering contrast. In the presence of another sublattice with non-vanishing scattering fluctuations Δb, both sublattices will contribute to the scattering. Since the condition (2.37) that a mean lattice as well as mean sublattices (denoted by s and s') exist holds for any inter- and intra-sublattice vector $R_l = R_s - R'_{s'}$,

$$\sum_{i,j} \langle c_s^i(R) c_{s'}^j(R') \rangle \langle u_l^{ij} \rangle = 0 . \tag{2.39}$$

If one considers, for instance, the cubic stabilized (Y-doped) zirconia $Zr(Y)O_{2-x}$, the cation contrast is rather small for neutrons, as well as for x-rays. Two non-negligible scattering contributions can be expected from the displacements between the oxygens and from those between oxygen and cations. Furthermore, (2.39) permits one to determine the displacements of the cations around the vacancies, $c^{vac} \langle u_l^{vacZr(Y)} \rangle = -c^O \langle u_l^{OZr(Y)} \rangle$.

In analogy with the discussion of the total scattering, where we distinguished the mean and fluctuating parts, we do so here again to separate the mean size effect from the part of the displacements which actually depends on the chemical nature of the different pairings of atoms involved. Therefore we rewrite (2.33), using the identity $c_m^i = c_m^i - c^i + c^i = \Delta c_m^i + c^i$. For the linear displacement scattering this reads as

$$\frac{d\sigma}{d\Omega}(Q)\Big|_{\substack{\text{lin.} \\ \text{disp.}}} = \sum_{i,j} b^i b^j \sum_l \langle (c^i c^j + + \Delta c_m^i c^j + c^i \Delta c_{m+l}^j + \Delta c_m^i \Delta c_{m+l}^j) $$
$$\times iQ \cdot (u_{m+l} - u_m)) \, e^{iQ \cdot R_l} . \tag{2.40}$$

Now we distinguish three terms in (2.40). The first term vanishes, since $\langle u_m \rangle = 0$, and reflects quite naturally that there is an average lattice. This vanishing contribution to the scattering would correspond to mere density correlations and would be proportional to $\langle b \rangle^2$.

The next two contributions will be due to terms linear in $\langle \Delta c_m^i u_{m+l} \rangle$ and $\langle \Delta c_{m+l}^i u_m \rangle$ and characterize the two-point correlation function of the mere size effect for the random alloy:

$$\frac{d\sigma}{d\Omega}(Q)\Big|_{\substack{\text{size} \\ \text{effect}}} = \sum_{\substack{i,j \\ i>j}} 2(b^i - b^j)\langle b \rangle \sum_l e^{iQ \cdot R_l} iQ \cdot \langle \Delta c_m^i u_{m+l} \rangle , \tag{2.41}$$

where we have used that $\sum_i \Delta c_i = 0$. This scattering term is proportional to $(b^i - b^j)\langle b \rangle$ as expected for a correlation between fluctuations in occupation and position.

We now turn to the last term and second contribution to the linear displacement scattering in (2.40) due to the specific chemical properties of the atomic pairs. It includes the remaining three- and four-point correlations. Using again that $\sum_i \Delta c_i = 0$, we can write it as

$$\frac{d\sigma}{d\Omega}(Q)_{\substack{\text{SRO}\\\text{disp.}}} = \sum_{\substack{i,j\\i>j}} |b^i - b^j|^2 \sum_l e^{iQ \cdot R_l} iQ \cdot \langle \Delta c_m^i \Delta c_{m+l}^j (u_m - u_{m+l}) \rangle. \quad (2.42)$$

As one can see, the sum of all terms beyond the two-point correlations is proportional to $|b^i - b^j|^2$. It can be separated from the short-range order scattering only by its Q-dependence. This term includes the cross-term of short-range order and size-effects and also the four-point correlation, which is due to a charge effect causing that the displacements between two atoms depend on their chemical nature. However, other effects may play a role as well, such as changes in the local force constants between the different kind of neighbor pairs.

For separating the displacement scattering in terms of size and charge effects one could also assume that

$$\langle u_l^{ij} \rangle \approx \langle u_{0,l}^i \rangle + \langle u_{0,l}^j \rangle + \langle \delta u_l^{ij} \rangle. \quad (2.43)$$

The displacements $\langle \delta u_l^{ij} \rangle$ can be related to the relative charges Δq^i of the atomic species in the alloy:

$$\langle \delta u_l^{ij} \rangle \propto \Delta q^i \Delta q^j, \quad (2.44)$$

because charge neutrality demands

$$\sum_i c^i \Delta q^i = 0. \quad (2.45)$$

Charge balance and the mean lattice condition assures that there are again only two independent variables describing the static displacements.

The physical origin of the "size" rests upon the weakly deformable part of the localized electronic shell of an atom, while the direct Coulomb interaction between ions, or atoms of an alloy with charge transfer between the species, causes an additional species-dependence of the displacements. If there was only a size effect in an AB alloy, where, say, A is bigger than B, one would observe $\langle u_l^{BB} \rangle < \langle u_l^{AB} \rangle < \langle u_l^{AA} \rangle$; on the other hand, if the Coulomb interaction was dominant, like and unlike neighbor pairs should have positive and negative displacements respectively, so that $\langle u_l^{AB} \rangle < 0 < \langle u_l^{AA} \rangle \approx \langle u_l^{BB} \rangle$.

Considering only size effects, for a random alloy, the displacement $u_{0,l}^i$ at a distance R_l due to the

Because of the elastic properties of a crystal the displacements $\langle u_{0,l}^i \rangle$ (the ones which cannot be distinguished with respect to the displaced atoms) are of long range, $\langle u_{0,l}^i \rangle \propto e_R/R^2$, and the Fourier transform will be $u_0^i(q) \propto e_q/q$. Close to the reciprocal lattice points G, the linear displacement scattering is proportional to $\pm Q/q$ and causes a comparatively strong diffuse scattering proportional to $\pm q^{-1}$ for $G \neq 0$ (divergent for an infinite crystal).

The range of the species-dependent part of the displacements $\langle \delta u_l^{ij} \rangle$ will be confined by the effective length in which the charge fluctuations are screened. Not only in metallic alloys, but also in disordered ionic systems, there is a mutual screening of exponential type because of the polarizibility and also because of a self-screening due to the short-range order itself. Hence, the range of $\langle \delta u_l^{ij} \rangle$ is expected to be smaller or comparable to that of the chemical short-range order.

The scattering contribution to (2.20) due to the second order displacement terms is

$$\frac{d\sigma}{d\Omega}(\boldsymbol{Q})_{\substack{\text{quad.}\\\text{disp.}}} = \sum_{i,j} b^i b^j \sum_l e^{i\boldsymbol{Q}\cdot\boldsymbol{R}_l} \langle c_m^i c_{m+l}^j (\boldsymbol{Q}\cdot\boldsymbol{u}_m)(\boldsymbol{Q}\cdot\boldsymbol{u}_{m+l})\rangle . \qquad (2.46)$$

At first, one may note that the Fourier transform of the brackets yields again a symmetric (even) function just as for the chemical short-range order, whereas its intensity varies as Q^2 which makes both contributions separable.

The number of distinct terms contributing to (2.46) is large. There are $\sum_i i$ (for the possible pairings of the species) times six non-equivalent combinations of the two projections of $\boldsymbol{u}$ onto the scattering vector $\boldsymbol{Q}$.

As a second-order term it is usually less important than the linear term. If one is interested primarily in the chemical short-range order it may be wise to explore the necessary volume of reciprocal space just around the origin, and the second-order terms may be negligibly small for most of the data. However, for large $\boldsymbol{Q}$ the higher moments will contribute significantly. In particular, in the long-wave limit, since $u_0^i(q) \propto e_q/q$, the quadratic term dominates the linear (antisymmetric) term, and leads to the so-called *Huang scattering* which is proportional to q^{-2} close to the Bragg peaks. This asymptotic behavior has often been observed, particularly for dilute defect systems. However, any deviations for very low q in particular could be indicative of whether the Gaussian distribution holds, at least for the small displacements between very distant atoms.

Equation (2.46) may be discussed in an analogous way to the linear displacement term by rewriting, for instance as above, the scattering in terms of the mean amplitude and the fluctuating parts, $b = \langle b \rangle + \Delta b$. This should result in three types of intensities: an average term regardless of the actual occupations,

$$\frac{d\sigma}{d\Omega}(\boldsymbol{Q})_{\substack{\text{1,quad.}\\\text{disp.}}} = \langle b \rangle^2 \sum_l e^{i\boldsymbol{Q}\cdot\boldsymbol{R}_l} \langle (\boldsymbol{Q}\cdot\boldsymbol{u}_m)(\boldsymbol{Q}\cdot\boldsymbol{u}_{m+l})\rangle , \qquad (2.47)$$

a second (quadratic) "size effect" term

$$\frac{d\sigma}{d\Omega}(\boldsymbol{Q})_{\substack{\text{2,quad.}\\\text{disp.}}} = \sum_{\substack{i,j\\i>j}} 2(b^i - b^j)\langle b \rangle$$

$$\times \sum_l e^{i\boldsymbol{Q}\cdot\boldsymbol{R}_l} \langle \Delta c_m^i (\boldsymbol{Q}\cdot\boldsymbol{u}_{m+l,m})(\boldsymbol{Q}\cdot\boldsymbol{u}_{m,m+l})\rangle , \qquad (2.48)$$

which is symmetric with respect to an exchange of i by j, and a third term

$$\frac{d\sigma}{d\Omega}(\boldsymbol{Q})_{\substack{3,\text{quad.}\\\text{disp.}}} = \sum_{\substack{i,j\\i>j}} |b^i - b^j|^2$$

$$\times \sum_l e^{i\boldsymbol{Q}\cdot\boldsymbol{R}_l}\langle \Delta c_m^i \Delta c_{m+l}^j (\boldsymbol{Q}\cdot\boldsymbol{u}_m)(\boldsymbol{Q}\cdot\boldsymbol{u}_{m+l})\rangle\,. \tag{2.49}$$

From this it follows that the scattering due to the autocorrelation ($\boldsymbol{R}_l = 0$) with the scattering weights $|\Delta b|^2$ cancels with the static Debye–Waller factor for the Laue scattering (compare (2.15) and (2.30)).

Because of the comparable energy of the neutrons it is quite simple to measure the phonon scattering separately. Considering the time-dependent correlations between distinct displacements – which are visible in the scattering by their projections on the scattering vector as $\langle(\boldsymbol{Q}\cdot\boldsymbol{u}_m(0))(\boldsymbol{Q}\cdot\boldsymbol{u}_{m+l}(t))\rangle$ – an energy resolution $\delta E < \hbar/t$ is sufficient to determine these separately from the static displacements. Since δE is typically large for x-ray scattering, one usually measures correlations at equal times $\langle(\boldsymbol{Q}\cdot\boldsymbol{u}_m(t))(\boldsymbol{Q}\cdot\boldsymbol{u}_{m+l}(t))\rangle$. Therefore, (2.46) includes the displacements due to the thermal vibrations in the crystal within the harmonic approximation (one-phonon scattering). Usually one calculates this background to separate it from the other scattering contributions. Of course, it may offer interesting information by itself. Before the time of neutron scattering, the first phonon dispersion curves were modeled according to the measured energy-integrated intensities. With synchrotron radiation the scattering contrast of near-neighbor elements can be diminished so that only the contribution of the mean lattice $\langle b\rangle$ survives. Part of the so-called 3λ method is to measure this background of thermal diffuse scattering (and the Huang scattering) for subtraction. In principle, the chemical short-range order may also leave its fingerprint on the thermal vibrations as well as on the static displacements; this is, however, usually neglected.

As already mentioned at the beginning of the discussion of the displacement scattering terms, the averages correspond to higher-order correlations, connecting two occupational variables with two positional variables in increasing powers. It has been pointed out by Dietrich and Fenzl [1.14] that the displacements in crystals also offer the opportunity of obtaining information about the *higher-order particle* correlations, i.e. triplets and so forth.

Therefore, the actual displacements $\boldsymbol{u}_m$ have to be substituted by the superposition of the displacement fields $\boldsymbol{u}_{m+l,m}^i$ of all the surrounding atoms specified by their type i,

$$\boldsymbol{u}_m = \sum_l \sum_i \Delta c_{m+l}^i \boldsymbol{u}_{m+l,m}\,. \tag{2.50}$$

Then, higher-order particle correlations appear in the configurational average, while inside this an average over the additional particles is taken with the weights of the displacement fields.

This linear superposition principle is an approximation, though admittedly a very useful one, and should hold for sufficiently small displacements so that the atoms still feel only the harmonic part of the potential. Furthermore, one has to assume that the coupling force constants are still globally valid and not modified along the specific "bonds" between the different atomic pairs. Using displacement fields also neglects displacements due to charge effects. It might well be that these approximations cause errors which are comparable to the effects of possibly inherent many-body interactions (which are expected to be of short range). Therefore this approach, though it is in principle very tempting, does not appear very promising for a quantitative determination of many-body correlations and interactions. However, at least qualitatively, the effect of higher order particle correlations seem to explain a typically observed variation of displacements between for first and second neighbors in ordering alloys. These are often of opposite sign and in the same order of magnitude, which can be understood from the superposition of the displacements between nearest neighbors only.

From a conceptual point of view a description in terms of the displacement fields is attractive, not only with respect to possible many-body correlations and the use of lattice properties but also to apply the so-called Kanzaki force models [2.16–2.20]. Krivoglaz [2.9] attempted to justify this approach further by arguing against "the unrealistic assumption that the distance between two atoms depends only on the nature of these atoms, since other atoms will exert no less influence". However, there is certainly nothing wrong with (2.20) which merely implies that mean values such as $\langle u_l^{ij} \rangle$ exist, although it is not a priori clear that the distribution of $\langle u_l^{ij} \rangle$ is unimodal.

The displacement fields are of long range because of the elastic properties of the crystal. The idea of the Kanzaki model is to substitute the long-range displacement fields $u_{m+l,m}^i$ by only short-range forces and the lattice Green's function. According to Hooke's law in linear response we can write

$$u_{\text{def}}(q) = \Phi(q)^{-1} f_{\text{def}}(q) \ . \tag{2.51}$$

The force-constant matrix Φ is related to the usual dynamical matrix by

$$\Phi(q) = M^{1/2} D(q) M^{1/2} \ , \tag{2.52}$$

and can be obtained from neutron data of the phonon dispersion. Fig. 2.1 gives a schematic view of a Kanzaki force model around a defect.

Equation (2.51) has been applied mostly to dilute defect systems, where Φ is taken as known for the ideal lattice (determined from measured phonon frequencies) and the forces $f_{\text{def}}(R_l)$ have to be consistent with the volume relaxation. In linear order, the trace of the dipole force tensor characterizes the macroscopic "strength" of the defect

$$P = \sum_l R_l \cdot f_{\text{def}}(R_l) = 3pV = 3B \Delta V \ , \tag{2.53}$$

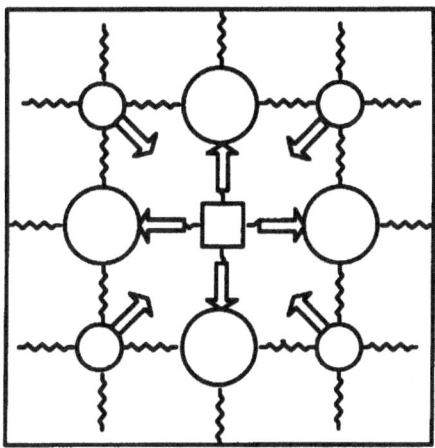

Fig. 2.1. Schematic representation of a Kanzaki force model (*arrows*) for the displacement field around a vacancy in an ionic crystal (e.g. FeO). The defect, here a cation vacancy, repels the neighbor anions and attracts next-neighbor cations because of the missing positive charges. The resulting displacement field is qualitatively similar to the force field for near distances but it is of long range due to the interatomic couplings

where B denotes the bulk modulus and the relaxation volume per defect ΔV is determined from the lattice parameter change with the defect concentration.

In *concentrated* alloys it does not make sense to distinguish between defects and host atoms. All constituents will have their typical displacement fields due to their specific sizes and ought to be considered as defects with respect to the mean lattice. Therefore, we may use (2.50) and by a Fourier transform the convolution on the right-hand side is factorized,

$$u^{(i)}(q) = \Delta c^i(q) u_0^{(i)}(q) , \tag{2.54}$$

$$f_{\mathrm{def}}^{(i)}(q) = \Delta c^i(q) f_0^{(i)}(q) , \tag{2.55}$$

to be inserted into (2.51); $f_0^{(i)}(q)$ is the force associated with a single defect relative to the perfect lattice. The atoms will deviate positively and negatively in size with respect to the mean lattice, and also the sum of forces will vanish for the configurational average $\sum_i \langle \Delta c_m^i f_{0,l}^{(i)} \rangle = 0$. For a concentrated binary system, for instance, $\Delta c_m^A = -\Delta c_m^B$, and Vegard's law [2.21] $(\partial a/\partial c =$ constant) implies $f_{0,l}^{(A)} = f_{0,l}^{(B)}$. Hence the actual force $\Delta c_m^i f_{0,l}^{(i)}$ exerted by A has the opposite sign to the force exerted by B and the strength of the defect force depends on the concentration according to the fluctuation, $\Delta c_m^A = c^B$ (if m is occupied by A) and $\Delta c_m^B = c^A$ (if m is occupied by B).

Considering of the displacements which arise because of the size effect only, the diffuse scattering may be evaluated as [2.9]

$$\frac{d\sigma}{d\Omega}(Q)_{\mathrm{diffuse}} = N|\Delta c(q)|^2 |\Delta b + i \bar{b} Q \cdot u_o(q)|^2 , \tag{2.56}$$

where Δb may denote either Δb^A or Δb^B.

If one now considers a random alloy, there will be only an incoherent overlap of all the displacement fields and the scattering will be proportional to that of the dilute case. However, correlated displacement fields tend to interfere destructively in the case of an ordering tendency, when the ordering wave vector coincides with the high-symmetry (Lifshitz) points of the reciprocal lattice. In contrast to this, one would expect in the case of clustering alloys a significant increase of the Huang scattering, which is nonlinear in c. Because of the elastic energy involved, the large-scale fluctuations (as $q \to 0$) of these displacements in a homogeneous solution will be less stable than a morphology of well-separated nuclei or regions of A- and B rich phases. For a fuller discussion the work of Khachaturyan [1.2] in particular is recommended.

In Sect. 4.3 an application of Kanzaki models to a concentrated pseudobinary system, $Fe_{1-x}O$, will be discussed. This application to an ionic system is unusual and has to be considered with caution, because the displacements are not typical for size but for charge effects, and therefore the forces and displacements depend on type of the displaced atoms. However, since cations and anions do not occupy a common sublattice, at least within a good approximation it is not necessary to consider this dependence explicitly. Although the range of the forces should still be finite, because of the screening effects, it may surprise how well the situation can be described by the very simple model, in which cation vacancies exert only two forces, one on the neighbored anion sites and and another force on the neighbored cation sites.

2.5 Data Analysis

There are two ways, or combinations of them, to approach the real structure behind the diffuse scattering pattern. One is a "brute force" method to establish from the scattering experiment the complete information about the available pair-correlation functions of the system under study using methods of Fourier analysis. By reverse Monte Carlo methods (see Chap. 3) a structure which is compatible with this information can be generated. In principle, the structure may also be generated directly from the measured intensities.

The alternative, especially for those cases where the information is rather incomplete, is to propose a plausible structural model and to test it on the available data. At least, scattering experiments could eventually falsify wrong models. The approach could be an analytic approximation, or, to avoid the complexity of the diffuse scattering terms, one could use computer models generated by Monte Carlo methods and a subsequent Fourier transform thereof to obtain a scattering pattern for comparison.

If a sufficiently large volume in reciprocal space has been measured, it is possible to determine the parameters describing the chemical short-range order and the lattice displacements in real space by Fourier analysis. A separation of the different Fourier series relies on the distinction in the series with respect to even ($S_{cc} \propto \cos \boldsymbol{Q} \cdot \boldsymbol{R}_l$, $S_{nn} \propto Q^2 \cos \boldsymbol{Q} \cdot \boldsymbol{R}_l$) and odd ($S_{cn} \propto \boldsymbol{Q} \cdot \boldsymbol{u} \sin \boldsymbol{Q} \cdot \boldsymbol{R}_l$) functions and their different dependences on the scattering vector $\boldsymbol{Q}$.

In order to make such a separation, the Borie-Sparks method [2.22, 2.23] compares data measured at points in reciprocal space which are equivalent by symmetry, followed by a Fourier transformation of various terms which separates short-range order and size-effect scattering.

Instead of this procedure and, in particular, if the measured data are not regularly dispersed in reciprocal space, the separation can be performed by a Fourier analysis using the standard *linear least-squares method* as introduced by Williams [2.24]. Using established, sophisticated algorithms (singular value decomposition [2.25]) one achieves the most reasonable and stable solutions. This method imposes the least restrictions on the experimental procedure. In particular, this type of analysis is convenient if multiple-detectors are used, allowing for a straightforward measurement of intensities in planes and volumes of the reciprocal space.

Quite often the quadratic displacement terms may be neglected in the analysis of elastic neutron data, since the thermal diffuse scattering is already separated experimentally. For instance, this is the case for all applications discussed in Chap. 4. Therefore the Huang scattering is not taken into consideration. However, this scattering contribution due to the static displacements in second order is typically only significant very close to the Bragg peaks.

In the case of x-ray data, the specific atomic Q-dependence of the x-ray scattering factors should be incorporated into the least square analysis. If the Q-dependence differs significantly for the alloying species, it can be used to separate the species-dependent displacements [2.26]. Therefore, a method based on a least squares analysis and known as the Georgopolous-Cohen procedure has been devised [2.27]. However, the price of this further information is that the number of parameters for only *one* distance in low-symmetry directions becomes as frighteningly large as twenty-five, requiring the measurement of the minimal irreducible volume with much repetition. Therefore, several thousand diffuse intensities are typically needed to determine about one hundred parameters describing the short-range order and lattice displacements.

The recent advances utilizing anomalous scattering and repeating the measurement with different scattering contrasts have considerably improved the reliability of the results. The 3λ method introduced recently by Ice et al. [2.6, 2.28] seems to be appealing in this context. For example, using the considerable change of the atomic scattering factors near resonance levels – typically around the K or L-edge of one component of the alloy – the scattering contrast (Δb) can be enhanced, which will reveal the short-range order in alloys of elements having a similar number of electrons. It also provides an elegant tool to remove efficiently the scattering contributions of second and higher order in $Q \cdot u$, which depend only on the average scattering amplitude $\langle b \rangle$, from those due to short-range order and linear displacement, which depend on Δb. Therefore, the idea is to perform an additional measurement with approximately zero scattering contrast by tuning the incoming wavelength (using a synchrotron source). [§] In addition, the 3λ method requires two further measurements with large but different contrast for an optimal separation of the scattering contributions due to short-range order and displacements.

The methods above were devised particularly for x-ray scattering (for further discussions see [1.12, 1.18, 2.29]). In contrast to neutron experiments, corrections have to be made for thermal diffuse scattering, Compton scattering, resonant Raman scattering close to the absorption edges and fluorescence. Energy analysis of the scattered x-rays permits one to remove at least a part of this background, e.g. the recent experiments at a synchrotron source of Ice et al [2.6, 2.28, 2.29].

For neutron investigations and for cases where the part of the diffuse intensity due to short-range order is of primary interest, it is much more favorable to investigate the scattering closer to the origin of reciprocal space, where displacement contributions are small. In view of the typically weak diffuse intensities (compared to those of the Bragg peaks) it is of course advantageous to use multiple detectors and, as mentioned above, a Fourier analysis by linear least-squares methods. Ideally, one would like to measure all elements of reciprocal space with equal weights and with the necessary repetition in order to make all Fourier coefficients independent of each other. With multiple detectors this would be rather inconvenient. The strategy for determining three-dimensional Fourier coefficients will be to measure as large a volume as possible with some generous overdetermination of the parameters. The least-squares procedure also provides standard deviations for short-range order and displacement parameters, including correlations among these parameters.

The simplest approach would be to calculate the diffuse scattering amplitude with respect to the ideal lattice rather than to the mean lattice, assuming a well-localized defect, which neglects the unavoidable long-range displacement fields of defects in solids. Therefore, replacing in (2.18) $\langle A(\boldsymbol{Q}) \rangle$ by $A_{\mathrm{ideal}}(\boldsymbol{Q})$, only the sum over a defect or indexdefect cluster "defect cluster" yields a form factor that matters for the diffuse scattering cross-section. This may include occupational defects, such as vacancies, interstitials or anti-site atoms, but also displaced atoms in the surroundings. To give examples, this type of model calculation has been used in studies of disorder in the non-stoichiometric oxides $UO_{2.13}$ by Goff et al [2.30] and $Mn_{0.94}O$ by Schuster et al [2.31]. Beyond its simplicity, a further advantage of such a model is that in the case of strong distortions, the phase factors containing the displacements need not be expanded, as required for the Fourier methods, but can be treated by nonlinear fit routines.

[§] Usually this is a good approximation, though one may note that this will not remove contributions from local phonon modes or short-range order effects on the phonons, which are also proportional to the scattering contrast. See also [1.13, 2.8, 2.29].

The so-called "micro-domain" models of Hashimoto [2.32] and Neder et al [2.33] were devised in the same spirit, and were applied to scattering data of Y-stabilized zirconia, for instance. In principle, these models allow one to include correlations among the defects or defect clusters. Models like these are appealing since they provide a simple visualization of coherently scattering defect arrangements. One has to realize, however, that imagining a specific and extended defect configuration, say a "cluster", already implies the need for correlation terms for the scattering. Therefore rather cumbersome "excluded volume terms" would be required to describe the tacitly assumed non-overlapping of any clusters. Analogous methods of data analysis can be found in small-angle scattering studies of heterogeneous structures, precipitates, etc., while the wide-angle diffuse scattering at larger Q has to account carefully for the displacements as well.

There has been a long-standing tradition in investigations of point defects and small clusters which can be treated within approximations of the dilute limit. For defect concentrations of about 1 at% or less, short-range order correlations become negligible and defects may be identified by the diffuse scattering due to their long-range displacement fields, in particular by the Huang scattering [2.34]. The Kanzaki force formalism as introduced above permits one to include the entire long-range displacement field by means of a very few force parameters which are further fixed by the macroscopic volume change due to the incorporation of defects. This enables one to distinguish between vacancies, possible interstitial positions, dumbbells and others. For a recent comprehensive review see Ehrhart [2.35], and for further details [1.1, 2.9, 2.36–2.38].

These model concepts may be developed further, as discussed in Chap. 4 for the example of non-stoichiometric FeO. In case of more concentrated solid solutions a treatment with respect to the mean lattice is feasible, which takes into account excluded volume effects and the entire displacement fields on the basis of the Kanzaki formalism. Nevertheless and despite the value and the success of this particular model, we note that the approximation of a defect cluster can be only a crude compromise to the real correlations in condensed matter.

The ultimate goal of scattering studies, however, is not to obtain a sufficient approximate mathematical description of the measured intensities, but to find a realistic model of the structure, which is compatible with the observed scattering. Therefore, one may use computer models which can be refined and compared via Fourier transformation with the measured intensities. Such an approach, as supported by Welberry in a recent review [1.20], could avoid the complexity of diffuse scattering formalisms. Of course, models can also be derived using the correlation parameters of short-range order and displacements that have been obtained from quantitative analysis based on the Fourier methods. In the latter case, the simulation is straightforward, and has been rather commonly applied since its proposal by Gehlen and Cohen [2.39]. The procedure can be regarded as a reverse Monte Carlo algorithm – a method better known from applications to liquid structures – and will be discussed in Chap. 3 in detail. Unfortunately, the information obtained from this approach, like any other, is not necessarily unique, since the scattering contains only direct information about the two-body correlations (see also Chap. 3).

2.6 Experimental Considerations

The substantial body of experimental results which will be discussed here has been explored by using cold and thermal neutrons. At first glance this might be surprising, since at a time when synchrotrons have been built all over the world, neutrons are no longer considered to be the optimal probe to investigate bulk structural properties in condensed matter on atomic length scales. Of course, there is some competition between both methods, and one has to find out which is the best-suited method for each problem. On the other hand, the methods themselves are unique and complementary in various respects.

If large single crystals are available, of the order of $1\,\mathrm{cm}^3$, there are some advantages in using neutrons for studies of disorder in alloys. In particular, cold and thermal neutrons have not only the appropriate wavelengths but also the energies to measure the disorder *and* to distinguish between static and dynamic origins (see also the recent review of Moss [1.21]). The separation between elastic and thermal diffuse scattering becomes particularly important for *in situ* high-temperature measurements (e.g. for the studies of FeO described in Chap. 4).

Neutron scattering amplitudes are very different for different elements but comparably large for light and heavy elements. They also vary between isotopes, which can be used with great advantage. Since some of these amplitudes even take negative values, it is possible to make "null-matrix" alloys to discriminate the scattering from the mean lattice (see, for instance, Sect. 4.1).

The differential cross sections for the diffuse scattering are typically weak compared to Bragg intensities, while after integration over the first Brillouin zone the intensities are quite comparable. Typically, only a modest resolution for the scattering vector Q is required in investigations of disorder phenomena, which helps to detect the weak diffuse scattering signals within reasonable times. In addition, special dedicated instruments use multidetector systems to collect the data efficiently in three dimensions of reciprocal space for Fourier analyses. Standard triple-axis instruments, which exist at nearly every research reactor, can be used for this purpose. However, there are more specialized machines for investigations of disordered systems. These use multiple-detectors and measure the time of flight of the neutrons for energy analysis – for instance, the instruments D7 at ILL-Grenoble and G4-4 at LLB-Saclay, or use multiple detectors plus many analyzers – such a multi-triple-axis machine is the "flat-cone" instrument at HMI-Berlin. The instrument DNS at the Forschungszentrum Jülich can operate in both modes. At the ISIS spallation source at the Rutherford Laboratory, the SXD instrument uses a pulsed, white beam to measure at *one* sample position a 3-D volume in reciprocal space with a 2-D multi-detector, but energy transfers can no longer be analyzed.

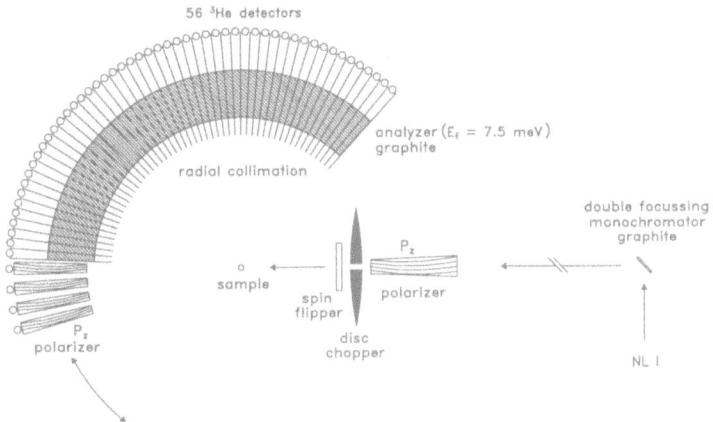

Fig. 2.2. DNS time-of-flight spectrometer in a schematic representation. The high flux at the expense of modest resolution properties is particularly convenient for diffuse scattering problems. Three options can be used: the usual time-of-flight mode, energy analysis by graphite crystals (for $E_f \approx 7.5$ meV), and (under development) polarizers preparing the incident beam and analyzing the scattered neutrons

Figure 2.2 shows the DNS (diffuse neutron scattering) spectrometer, which is located at the Jülich research reactor and uses neutrons from a hydrogen cold source. The energy analysis can be performed by a conventional time-of-flight method with a resolution between 0.25 meV and 1 meV. For example, a time-of-flight spectrum is displayed in Fig. 2.3.

There is a significant variation with scattering vector Q (or angle θ) of the diffuse elastic intensities, which can be seen at fixed time of flight near channel 90. Because of the specific sample orientation, the strong 200 Bragg reflection appears as well. Note that the frequent inelastic scattering events due to (multi-)phonons can be distinguished well from the diffuse elastic signal of interest. The measurement has been performed on an $Fe_{1-x}O$ single crystal in equilibrium with a surrounding CO/CO_2 gas mixture which is used to determine the oxygen partial pressure at high temperatures. Note that these raw data still contain all background contributions, for instance from the furnace. The diffuse elastic intensities can be measured in a reciprocal lattice plane by rotating the sample. Three-dimensional data can be obtained by an reorienting the single crystal or lifting the detectors out of

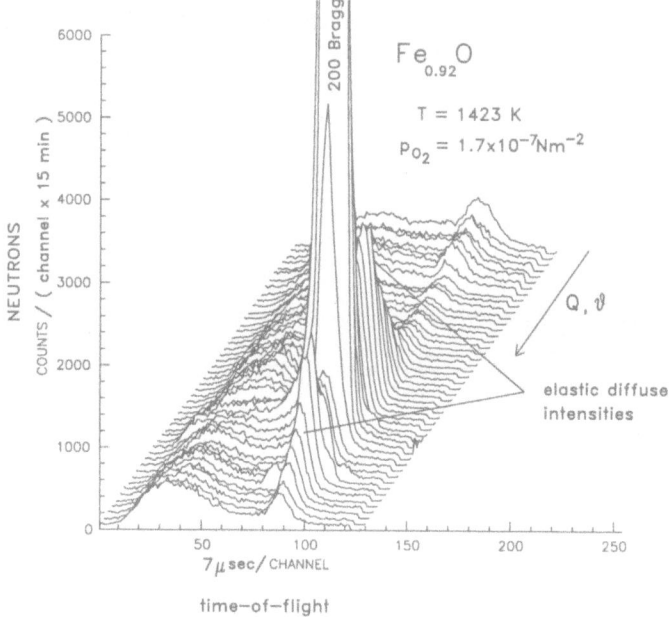

Fig. 2.3. Time-of-flight spectrum as measured for a $Fe_{1-x}O$ single crystal exposed to a reducing atmosphere at high temperatures (raw data). The diffuse elastic intensities, which can be seen at fixed time of flight near channel 90, vary significantly versus scattering vector Q. In addition, there is inelastic (phonon) scattering as well as intense scattering from the 200 Bragg reflection

the horizontal plane.

Alternatively, a multi-triple-axis mode can be used, and without the need for a chopper, the scattered neutrons can be analyzed by graphite crystals in front of the detectors, in order to collect only the elastically scattered neutrons. The third option using polarizers for incident and scattered neutrons is a current now development, and has not yet been used in the applications discussed in Chap. 4. This facility will certainly be of interest for determining spin–spin correlations in magnetic alloys. Another important application is to distinguish between the coherent and spin-incoherent scattering contributions [2.40, 2.41]. (Interested users should also see [2.42] on the internet.)

The x-ray radiation of synchrotron storage rings has other important advantages. The brilliance of the intense radiation is particularly useful for investigations of small single crystals. The scattering contrast can be easily tuned by a few electron units for x-ray energies close to an absorption edge of an element, which enables, for instance, the determination of species-dependent displacements in alloys, e.g. [2.28, 2.43]. Studies of near-surface order–disorder phenomena can be nicely explored by the scattering of the total reflected beam under grazing incidence (e.g. Dosch [2.44]). Recently, at

HASYLAB in Hamburg, R. Bouchard and H. F. Poulsen have developed a beamline for *hard* x-rays (of the order of 100 keV), which has been used in studies of oxygen vacancy disorder in high-T_c superconductors by Schlegel et al. [2.45]. This promising new technique opens new fields of applications. One advantage of the weak absorption of the hard x-rays is that the scattering volume is greatly increased. A second advantage of this property is that there are fewer restrictions on complicated sample surroundings like furnaces and cryostats, in comparison with the conventional x-ray scattering technique.

3. Monte Carlo Simulations of Alloys

It is of great fundamental interest to understand what stabilizes a particular alloy structure at a given concentration and temperature, and practical implications are evident for metallurgical problems. The well-known Ising model [3.1] has been widely used with continuing success to treat the configurational statistics of alloys on a microscopic basis using phenomenological atomic interaction parameters (see de Fontaine [1.1]). Diffuse scattering experiments offer direct experimental access to the atomic interactions. One may easily anticipate that for real alloys the true interactions will be more complex, of course, than a simple nearest-neighbor interaction model. First, a brief background about Ising models will be given in Sect. 3.1. The thermal equilibrium properties of alloys may be calculated by mean-field approximations, the Bragg–Williams model or, with much improved accuracy, by its natural extension, the various cluster models used in the cluster variation method (CVM) [1.1,3.2–3.5]. Monte Carlo simulations have been established as a useful tool to study order–disorder phenomena in alloys. There are a series of introductory textbooks and reviews of the Monte Carlo method, e.g. Binder [3.6] and [3.7–3.15]. The main advantages of the Monte Carlo method are that it can provide the most accurate results, and that it is straightforward to apply the method to complicated, realistic interaction models, which, for instance, may be of a longer range than the CVM is capable of handling. A brief description is given in Sect. 3.2. In this review, it will be of particular interest to discuss further the merits of Monte Carlo methods in relation to determinations of the structures (Sect. 3.3, Reverse Monte Carlo Method) and the atomic interactions (Sect. 3.4, Inverse Monte Carlo Method) from the measured pair correlations by scattering experiments.

3.1 Ising Models

The configurational energy of an alloy system may be described in a very simplified approach as

$$\mathcal{H} = \frac{1}{2} \sum_{m,n} \sum_{i,j} V^{ij}(\boldsymbol{R}_m - \boldsymbol{R}_n) c^i(\boldsymbol{R}_m) c^j(\boldsymbol{R}_n) - \sum_m \sum_i \mu^i c^i(\boldsymbol{R}_m) , \quad (3.1)$$

where m and n label lattice sites and i and j denote the species. Equation (3.1) implies a restriction to only pairwise interactions $V^{ij}(\boldsymbol{R}_m - \boldsymbol{R}_n)$ between the atoms.

For a binary AB alloy, one may introduce spin variables $s_m = 2c_m - 1 = \pm 1$ to denote the possible occupations by A and B atoms. Therefore, the binary alloy as well as the unary lattice gas can be mapped on the Ising model

$$\mathcal{H} = -\frac{1}{2} \sum_{m,n} J_{mn} s_m s_n - h \sum_m s_m \, , \tag{3.2}$$

apart from an irrelevant constant energy term. The pairwise interactions are determined by a linear combination of V^{ij}_{mn},

$$J_{mn} = -\frac{1}{2} V_{mn} = -\frac{1}{4}(V^{AA}_{mn} + V^{BB}_{mn} - 2V^{AB}_{mn}) \, . \tag{3.3}$$

The assumption that the difference of the chemical potentials is independent of the actual sites m is equivalent to an external magnetic field h acting on the system. This effective "magnetic" field reads as

$$2h = - \sum_{m(\neq n)} \{V^{AA}_{mn} - V^{BB}_{mn}\} + \Delta\mu \, . \tag{3.4}$$

From (3.3) and (3.4) it can be seen that differences between V^{AA}_{mn} and V^{BB}_{mn} do not lead to an asymmetry of the bulk configurational energies with respect to the equiatomic composition. However, at surfaces only a part of the interactions contributes to the sum in (3.4) which modifies the effective external field near the surface. Hence differences between V^{AA}_{mn} and V^{BB}_{mn} are expected to cause observable near-surface enrichments of one of the alloy components. In addition to the bulk field, (3.4), one has to account for these differences for near-surface sites m' by a term $-\sum_{m'} h_{m'} s_{m'}$. For example, for the fcc structure and only nearest-neighbor interactions we have $h' = V^{AA}_{mn} - V^{BB}_{mn}$, acting only on the surface layer.

Instead of (3.1), more generally one may include many-body interactions

$$\mathcal{H} - \sum_\alpha V_\alpha \upsilon_\alpha \, , \tag{3.5}$$

where $\sigma_\alpha = \prod_i^\alpha c^i_m$ denotes the occupation variable for a specific "figure" α, and V_α would be the related interaction energy. Figures could be, for instance, pairs, triplets or other atomic arrangements. One may distinguish between interaction models in which such figures may overlap and those in which this is excluded. The usual Ising-type models belong to the first case and generate correlation functions (Ornstein–Zernike) of more distant range than the interaction range, while this would not be the case for non-overlapping interaction models, which could model properties of systems with a molecular type of binding. Ising-type models which include many-body terms can be well approximated by effective pairwise interactions, as will be discussed in more detail below.

One may distinguish between two kinds of typical applications of Ising-type models: the first kind, using the most simplified models to study the generic behavior and universal, critical phenomena, and the second kind, which attempts to mimic the real properties of a specific alloy by incorporating more and more complex details in the interaction model. Even for the very simplified models there remain problems which are difficult and tedious to solve by Monte Carlo methods, despite all currently available computing resources (see, for instance, Sect. 4.4). As an alloy represents a complex system with many degrees of freedom, one has to realize that (3.2) imposes severe restrictions which must lead to contradictions with observed alloy properties. For instance, (3.2) leads only to coherent phase diagrams which are symmetric with respect to the concentration $c = 0.5$ or the field $h = 0$. In principle, one may account for such realistic demands by introducing concentration-dependent interactions and or many-body interactions, as in (3.5). In some alloys, magnetism plays an important role in the chemical ordering (e.g. Fe–Al, discussed Chap. 4), which requires an additional interaction term for the spin variables. Including positional degrees of freedom is feasible as well. Therefore, the ingredients can be inferred from experiments. Phonon measurements provide information about the force-constant matrix, while Kanzaki forces and the gradient of the effective interaction potential can be related to the measured displacements in alloys [3.16].

The naive expectation that simple pair interactions, on a rigid lattice and independent of concentration, are able to describe real alloys would also contradict Hume-Rothery's empirical rules [3.17]. These rules predict what happens if two metals are alloyed: (i) the alloy's electron-per-atom ratio e/a is correlated with the alloy's tendency to order or to cluster; (ii) the "size-effect" causes insolubility if the sizes differ more than 15%, and if the difference is less, the "big" and "small" atoms tend to order; (iii) there is an electronegativity effect (charge transfer) on the alloy phase stabilities arising from the chemical differences between the atoms.

Modern self-consistent theories can provide us with estimates of the effective interactions. Car and Parinello [3.18] have set a new milestone in the atomic-scale simulation of real matter. Their simulated annealing method treats the model system at finite temperature in a quantum mechanical approach, indicating that the time has come when phenomenological models, like the Lennard–Jones potential or Ising models, can in principle be avoided. However, this task is difficult and the most sophisticated techniques developed so far [3.19, 3.20] do not seem to work well for transition metal alloys. Therefore, to discuss the relevant states at finite temperature, in many cases Ising models are and will be useful or necessary. Nonetheless the seed has grown, and there are recent examples of including the electronic properties directly in Monte Carlo simulations (for a further introduction to simulation techniques the interested reader is also referred to a recent book of

Stoltze [3.21], and for calculations of lattice displacements in alloys see, for example, [3.22]).

In some instances such theoretical predictions appear to be sufficiently accurate, while in some other cases, despite all efforts and progress, there remain severe discrepancies between the predicted and observed alloy ordering properties. One needs to aim at a precision of about 1 meV for meaningful comparisons, which is very small compared to typical total binding energies.

Current theories pursue two different approaches to deduce such interactions from calculated electronic band structures. The first is the perturbative approach, which uses as a reference medium either an averaged disordered structure, in the coherent potential approximation (CPA), or the pure and the ordered structures (e.g. [3.23,3.24]). The second approach can be characterized as a matching scheme, first proposed by Connolly and Williams [3.25]. The total energies are calculated for a (large) set of different ordered structures, which one tries to map onto a compatible Ising-type Hamiltonian. Originally, only nearest-neighbor interactions were determined; for recent extensions to a determination of more realistic interaction models of longer range and containing multi-site terms, the reader is referred to a review by Zunger [3.26]. For the perturbative approach the results for the individual interaction terms remain unaltered within the approximations made, although essential interaction terms may have been disregarded. On the other hand, the matching schemes can provide only the true interactions, if the a priori choice of the Hamiltonian is an adequate choice and is a sufficient and complete basis.

As far as we are concerned, with metallic alloys, some justification can be given for Ising-type interactions. According to Pauling [3.27] "the metallic bond is closely related to the ordinary covalent or electron pair bond ... with the bonds resonating among the available positions in the usual case that the number of positions exceeds the number of bonds." This also points to the necessity of taking into account many-body terms as argued by Heine and Hafner [3.28], which can be done in a generalized Ising-type Hamiltonian. The lowering of the bonding energy due to the resonance effect is proportional to the square root of the coordination number, as derived by Heine [3.29] in tight-binding theory.

In contrast to metallic bonding, an Ising-type description is clearly not adequate for the molecular type of bonding, where the number of bonds is identical to the coordination number. In such a case one needs to distinguish between coordinated neighboring atoms and unbonded atoms which, nevertheless, may be accidentally just as close.

There is obviously a strong interest in checking the first-principles calculations by comparison with the effective atomic pair interactions as determined in a unique experimental way from the diffuse scattering.

A substantial body of diffuse scattering data on equilibrium short-range order has been analyzed in terms of effective pair interactions, since a mean-field approximation such as the Krivoglaz–Clapp–Moss (KCM) formula [2.9, 3.30, 3.31] has become popular:

$$\alpha(q) = \left(1 + 2c_A c_B \frac{V(q)}{k_B T}\right)^{-1}. \tag{3.6}$$

Equation (3.6) relates the Fourier transform of the short-range order to the Fourier transform of the pair interactions. The equation is expected to apply in the high-temperature limit, $V \ll k_B T$.

In addition to the determination of the interactions, the KCM formula permits one to draw a variety of conclusions: it provides the critical transition temperature $T_c = -2c_A c_B V(q_c)/k_B$; it predicts the spinodal line as the denominator vanishes; it says that curves (c, T) of equal short-range order exist; and it contains the Curie–Weiss law (for $q = q_c$) giving the mean-field critical exponent for the temperature dependence of the susceptibility, $\gamma = 1$.

The validity of (3.6) and recent improvements by Tokar, Masanskii and coworkers [3.32–3.34] have been discussed in a recent review by Reinhard and Moss [3.35]. This aspect and the general mean-field predictions will also be considered below, when the results are described for the alloys CuZn and NiCr in Chap. 4.

However, there is no need any more for such approximations to determine atomic interaction parameters. For better accuracy one should prefer the inverse Monte Carlo method (discussed below) or, at least, the inverse cluster variation method [3.5].

For a given set of interactions, valuable information on the ordering properties of an alloy can be concluded from a comparative analysis of ordered ground states. The possible ordered structures for various compositions can be categorized by structures belonging to typical ordering families (see [1.1, 3.36, 3.37]). This can be done without using the Monte Carlo method. Actually, such an analysis is a very useful prerequisite for Monte Carlo simulations of phase diagrams, since thereby one may define the relevant order parameters.

3.2 Metropolis Monte Carlo Method

There are complex ordering phenomena that may arise from the collective behavior of a many-particle system. With increasing number of atoms N, the possible configurations x increase as 2^N, and except for very small systems containing only a few atoms, and some trivial cases (i.e. for $T = 0$ or ∞), exact analytical solutions do not exist for the configurational equilibrium properties of binary alloys. The aim of the Monte Carlo (MC) method, following the idea of importance sampling of Metropolis et al. [3.38], is to generate the most likely states and configurations, and hence to obtain appropriate thermal averages of interesting observables. The probability of finding the system in a specific configuration, denoted by x, follows the Boltzmann distribution

$$P(x) = \frac{1}{Z} e^{-\beta \mathcal{H}(x)}, \quad \text{with} \quad \beta = (k_{\mathrm{B}} T)^{-1}, \tag{3.7}$$

where Z is the partition sum

$$Z = \sum_x e^{-\beta \mathcal{H}(x)}.$$

Since this distribution becomes sharply peaked close to equilibrium configurations x_{eq}, a simple MC sampling of randomly selected configurations will not yield appropriate thermal averages. The Monte Carlo method is essentially based on the principle of detailed balance,

$$P(x)W(x \to x') = P(x')W(x' \to x). \tag{3.8}$$

The idea of *importance sampling* is that a transition probability W, which is consistent with (3.8), does not only describe the thermodynamic equilibrium. The application of W to an arbitrary initial configuration generates a sequence of configurations via a Markovian process such that ultimately the relevant phase space is preferentially sampled. This principle properly prescribes a path to thermal equilibrium, as ergodicity can be presumed.

In practice, one first has to construct a lattice with periodic boundary conditions. It is particularly efficient for the determination of equilibrium properties to use the (semi-)grand canonical ensemble $(\Delta\mu, T)$ (see Binder [3.6]). This avoids degenerate two-phase equilibria and implies the least constraints on the relaxation process. However, for alloys this means a rather artificial Monte Carlo step, namely an exchange of atoms at one site ("spin-flip", $s_i \to -s_i$). One selects randomly a lattice site and calculates the energy change due to the "spin-flip". This attempted Monte Carlo step will be accepted if the transition probability

$$W = e^{-\beta(\mathcal{H}_i^{\mathrm{final}} - \mathcal{H}_i^{\mathrm{initial}})}$$

is larger than a random number $\eta \in (0, 1)$.

Generally, such MC step needs to be repeated often for all sites, in order to relax the whole system towards equilibrium, and by continuing sampling one can obtain many configurations to determine averages of composition, energy and other quantities for each T and h.

There are alternative choices for the transition probability. For instance, one may use $W' = W/(1 + W)$ (kinetic Ising model) [3.39, 3.40]. In the case of more than two different states – there are three for a ternary alloy or a binary alloy with one magnetic component (e.g. FeAl) – the so-called *heat-bath algorithm* [3.41, 3.42] is advantageous. Here only the final energies are used for the transition probabilities W_i, which have to be normalized by the sum of the transition probabilities to all possible final states.

If one is really interested in the kinetics of ordering or disordering in alloys, the canonical ensemble, which fixes the number of atoms of type A and B, appears to be a more natural choice. Nevertheless, a Monte Carlo step, which would typically be an exchange of nearest-neighbor atoms, is still unrealistic, and results should be compared with experimental observations with some caution. For more realistic studies of ordering kinetics one may introduce vacancies. Since they typically occur in small numbers, this would reduce the relaxation towards equilibrium by orders of magnitude, if one tried to use vectorizing algorithms, for instance on a Cray computer. However, with a single workstation processor assigned to a single vacancy one can achieve efficient simulations. The attention of the interested reader is drawn to the accurate and large database of binding energies between vacancies and metal impurities as well as host metals, obtained in first-principles calculations by Klemradt et al. [3.43, 3.44].

There is a considerable freedom to design the actual Monte Carlo step. This is also noteworthy in relation to applications within the reverse and inverse Monte Carlo methods, as proposed below. For instance, to compete with critical slowing down "flipping" of whole clusters has been tried [3.45], and other MC steps can be imagined.

The Monte Carlo method itself is very simple and with increasing effort the results may be as accurate as desired. However, the quality of the results depends on a careful analysis of the statistical *and* systematic errors. The Monte Carlo method as outlined above readily permits one to calculate, for instance, pair-correlations for comparison with scattering experiments. Close to continuous phase transitions, however, the correlation length and the correlation time diverge. Therefore, the finite size effects of the models need to be analyzed and long Monte Carlo runs are required. According to finite size scaling theory [3.46–3.52] we expect that instead of a real singularity the susceptibility $k_B T \chi(T) = N(\langle s^2 \rangle - \langle s \rangle^2)$ in a finite system exhibits, rather, a rounded maximum at $T_c(L)$ scaling with the linear dimensions L as $L^{\gamma/\nu}$. A plot of $T_c(L)$ versus $L^{-1/\nu}$ yields the T_c for an infinite system. Difficulties may arise if the exponent ν is not known. Analyses of the finite size dependence of the fourth-order cumulant $\tilde{U} = 1 - \langle s^4 \rangle / 3 \langle s^2 \rangle^2$ of the order parameter $\langle s \rangle$ is a quite elegant method to locate a second-order transition. Binder [3.50] has shown that the fourth-order cumulant of the order parameter depends on T and L but crosses at T_c for different values, where the fix-point $\tilde{U}^*(T_c)$ is a *universal* quantity. At $T = 0$, $\tilde{U} = 2/3$, and for $T \ll T_c$ the "magnetization" – or concentration in the alloy terminology – takes a Gaussian distribution, so that this higher-order cumulant tends to zero. Independently the exponent ν is determined from the slope

$$\frac{\partial \tilde{U}_{L'}}{\partial \tilde{U}_L}\Big|_{\tilde{U}^*} = (\frac{L'}{L})^{1/\nu} . \tag{3.9}$$

At a first-order transition, hysteresis effects may become severe, as it may be in nature. The transition can be accurately determined by a comparison of the free energies. Sometimes it has been argued that the Monte Carlo method has the disadvantage that the entropy and the free energy are not obtained like other thermal averages in a single MC run. One solution can be found by using standard thermodynamic relations, and by integration along paths in the variables T and h the changes in the free energy can be calculated from known reference states at $T = 0$ or infinity [3.53]. This method can yield very precise results, as shown for the first-order bulk transition temperature of the CuAu model discussed in Sect. 4.4.[§] Lim et al. [3.54] have described how the partial and integral quantities can also be obtained from single Monte Carlo runs for both the canonical and grand canonical ensembles. Hüller [3.55] proposed the use of the micro-canonical ensemble because it provides an easy access to the entropy in a single MC run.

Finally, we note a few technical points which may be of importance.

As pointed out in [3.15], high flexibility and performance of a program can be achieved if the geometry is treated separately. Therefore, the lattice is mapped onto a one-dimensional array, a chain defined by a regular walk along all sites. This procedure leads to helical periodic boundary conditions, which are intrinsically fulfilled within the chain, while at its ends an identical copy has to be added. It also facilitates our ability to vectorize a program on supercomputers. However, one should note that these boundary conditions cannot fit any possible ordering structure, e.g. the Cu_3Au structure. This can be accomplished by first distinguishing between the various sublattices, and using multiple parallel linear chains. Nevertheless, for specific problems, as described for instance in Sect. 4.4 for the near-surface ordering phenomena, it is necessary (at least partially) to return to the usual periodic boundary conditions.

Multi-spin coding techniques promise a considerable further efficiency, which, however, has to be paid for with some loss in flexibility. This technique is an optimal choice only for very simple interaction models. Here, the occupation variables are stored in the bits of an integer word, and logical operations allow for parallel computations [3.56, 3.57]. Thus, 335 million MC steps per second have been achieved on a single processor of a Cray-YMP [3.58].

Before starting the first MC simulation, particular attention has to be paid to the quality of the pseudo-random numbers. Random number generators based on the algorithm of Kirkpatrick and Stoll [3.59] are frequently used for Monte Carlo simulations. However, in principle the quality of a random number generator has to be checked every time with respect to its specific application. For discussions of appropriate random number generators we refer to the literature, e.g. [3.59–3.64].

[§] Taking a large number of Monte Carlo estimates along these paths is not really a disadvantage, since by integration all values for the internal energy (or "magnetization") contribute to reduce the statistical error of the determined free energy.

3.3 Reverse Monte Carlo Method

The Gehlen–Cohen procedure [2.39] uses the composition and short-range order parameters (point and pair correlation functions) to model the short-range order as observed in a diffuse scattering experiment. In the first of such modeling attempts Gehlen and Cohen used an fcc alloy model of 8000 atoms and the first six measured short-range order parameters.

Apart from the constraints due to composition and short-range order, one may expect that the entropy is maximized and a configuration is realized which is representative for the equilibrium state of the sample. If this is true, and this is the idea of the procedure, it offers a key to analyzing higher-order correlations although only pair correlations are usually observable in scattering experiments. This justifies the analysis of simulated structures in terms of possibly significant configurations. For example, the nearest neighborhood of an atom in an fcc alloy has twelve sites with 2^{12} different configurational patterns, which, however, can be reduced by the symmetry properties to only 144 so-called "Clapp-configurations" [3.65].

Of course, with the increasing computer power available nowadays, there is no need for the above-mentioned restrictions. The measured short-range order can be simulated with sufficient accuracy in much larger models and also within the whole range of significant correlations.

The algorithm is to accept any exchange of two unlike atoms if it only improves the simulated short-range order until it agrees with observation. One may note that a largely similar procedure has been applied to the simulation of liquids by Renniger et al [3.66]. Since the simulation with continuous degrees of freedom costs severely more computational effort, only a model of a few hundreds of atoms has been studied in this case.

One may suspect [3.67,3.68] that using only converging refinements could lead to incorrect results, which depend, for instance, on the initial state. In contrast to the standard MC method, in which one samples different configurations after relaxation close to equilibrium for reliable averages, this procedure freezes in the once-relaxed configuration. It should be noted that in this case a strictly convergent procedure does not necessarily lead to wrong results, provided that the systems have been chosen to be sufficiently large (this was checked for systems having 200 000 atoms). For ergodic systems the configurational averages should in principle be sufficient, although time averages are usually performed in standard Monte Carlo simulations. To overcome such problems, one could, for instance, allow all exchanges which are compatible with the short-range order within their error bars. More precisely, as done in the *reverse Monte Carlo method*, and recommended here as well, one should choose a proper weight for the acceptance of an exchange. In any case, the system size should at least be large compared with the correlation length of the short-range order.

The purpose of the reverse Monte Carlo method as proposed by McGreevy and Pusztai [3.68] is again structural modeling based on the point (densities, composition) and pair correlations obtained from scattering experiments. It was originally designed for, and has been frequently applied to, liquids, molecular liquids and glasses. The transition probability $w = w(\text{initial} \to \text{final})$ can be chosen as

$$w = \exp\left[-(\chi_f^2 - \chi_i^2)/2\right] , \qquad (3.10)$$

with

$$\chi^2 = \sum (\alpha_k^{\text{model}} - \alpha_k^{\text{exp}})^2/\sigma_k^2 ,$$

where σ denotes the standard deviation. Then, if $\chi_f^2 < \chi_i^2$ the change is accepted, and if $\chi_f^2 > \chi_i^2$ the change is accepted with probability w. The refinement of the observables α can be made with respect to the correlation functions or to measured intensities. In analogy to the standard MC method this transition probability w replaces the usual Boltzmann weight $w = \exp\left[-(\mathcal{H}_f - \mathcal{H}_i)/k_B T\right]$.

One may further note the analogy of σ_k^2 and T. The experimental accuracy not only determines the demands on the accuracy of the correlations in the model but furthermore confines the possible fluctuations. In particular, if a measurement has yielded very precise results for the α_k, the structural simulation by the reverse MC method starts with almost only converging moves being accepted, and the equilibrated structure, which will be found after a long simulation time, could even be frozen in, because any further change costs too large a fluctuation in $\Delta\alpha_k$. In this case the reverse MC method reduces to the criticized algorithm of Renniger [3.66] or Gehlen–Cohen [2.39] of only convergent moves. The analogue would be a standard MC simulation in the limit of zero temperature. Typically, at low temperatures the systems do not behave ergodically in practical MC simulations. Trapped in a metastable state defined by a local minimum of the free energy, it can take forever to reach the real global minimum. Therefore, different initial conditions usually need to be tested and a careful ground state analysis needs to be done. If one can expect only one minimum of the free energy, as for the simple liquid state, the reverse MC method (even for arbitrary small σ_k^2) should yield a proper representative configuration. In lattice gas models with a hard core repulsion only, there is a finite solubility at zero temperatures and thus also the entropy remains finite. This is not in conflict with thermodynamics since the configurational changes do not cost any internal energy. (For instance, all Ising Hamiltonians with interactions of only antiferromagnetic type, $J < 0$, have this property. For an fcc structure and J_1 only, we have the simplest model for the Cu-Au type of alloys, which yields not only the correct ordered structures but also the large solubilities of the disordered phase which are found in a number of alloy systems forming ordered compounds.) In an MC simulation, therefore, these systems do not freezein but the acceptance rate for MC steps approaches a constant as $T \to 0$, which depends also on the choice of the particular MC step, whether it is a "spin-flip" algorithm (Glauber), an exchange of two neighbors (Kawasaki) or of any two unlike atoms, or any other more complicated configurational change. Equivalently, the finite entropy of a disordered system ensures that freezing does not take place in the reverse MC simulation. However, in the case of simulations with continuous degrees of freedom, it may require a real art of simulation to find particularly suitable choices of attempted configurational changes, since the movements in dense liquids can be highly correlated, for instance.

Finally, it is worth saying that such structural simulations are also important because they offer a first test of the experimental results and can at least prove whether the observation in a scattering experiment corresponds to a possible physical configuration at all. Since the method can be used for liquids, there are of course no real problems in including lattice displacements in the simulation of structures. Since the displacements have been separated and determined in addition to short-range order parameters in very few systems, such efforts are just beginning to be made in current investigations.

3.4 Inverse Monte Carlo Method

With the *inverse Monte Carlo method* [3.69], as introduced by Gerold and Kern, it is possible to determine effective pair interactions which are consistent with the simulated structure as obtained by the procedure mentioned above. Thus, the reverse Monte Carlo algorithm can be considered as the first step within the inverse Monte Carlo procedure. Assuming that the measured and simulated short-range order describes a configuration of thermal equilibrium, one can apply the principle of detailed balance. Essentially one has to establish and to solve numerically a set of l nonlinear equations:

$$\sum_k \Delta p_{kl} w_k = 0 , \tag{3.11}$$

where Δp_{kl} is the change in the number of "bonds" of type l associated with the interaction energy V_l, $w_k(\Delta \mathcal{H}(V))$ is the attendant transition probability and

$$\Delta \mathcal{H}_k = \sum_l (p_{kl}^{\text{final}} - p_{kl}^{\text{initial}}) V_l \tag{3.12}$$

is the change of the configurational energy for a particular attempted fluctuation k. Equation (3.11) only holds for the average over a large number of fluctuations. The fluctuations only test the local minimum of the free energy and are not actually performed to keep the system in or near equilibrium.

Alternatively, one can also use the standard MC method and refine the interaction parameters in order to model the short-range order as done by Livet [3.70], which yields equivalent results [3.71].

There are further possible modifications and alternatives to the original inverse Monte Carlo method of Gerold and Kern. In combination with a reverse MC method, which produces also a time average of representative configurations, instead of the test fluctuations one could also use the configurational changes for all attempted fluctuations as soon as the system has relaxed sufficiently close to equilibrium and relevant states of the thermal equilibrium are sampled.

There have been numerous applications (e.g. [2.43, 3.35, 3.69, 3.71, 3.73, 3.75–3.77]) of the inverse Monte Carlo method to the short-range order in binary alloys since the method was proposed. Despite the larger number of diffraction studies on liquids or amorphous systems there has been only one example of a Lennard–Jones model liquid, studied by Ostheimer and Bertagnolli [3.78] which demonstrates its applicability to such non-crystalline cases. In practice, it seems to be necessary, but also possible, however, to improve the techniques and the numerical effort in order to achieve the desirable quantitative precision for systems having continuous degrees of freedom.

Unlike the MC method, which determines in a straightforward manner thermal averages, for instance, of correlation functions, the inverse method requires assumptions about the answer, which means that a priori one has to make a choice of the interaction model. Although in the very first applications [3.72] of the inverse MC method it was attempted also to determine many-body interactions, in subsequent applications it was found that the correlations among the interaction parameters became close to one [3.74], indicating that there is only a unique solution if one is restricted to an effective pairwise interaction model. The reasons will be discussed further below. Such a solution will be true, provided that the interaction model is also complete. For instance, in magnetic alloys there is a significant influence of magnetic interactions on the chemical order [3.75], which has to be taken into account. Bieber and Gautier [3.79] showed (by calculations in the tight-binding coherent potential approximation) that for transition metal alloys the pair interaction should be rather dominant compared with higher-order cluster interactions. Defining all interactions at a given range, this yields an upper limit for the higher-order cluster interactions, as follows from the discussion above. The accuracy to which interactions can be determined depends on that of the measured short-range order, and further on the numerical effort. However, it is obvious that an estimate is only possible within the range of a few $k_B T$. If the values of the effective interactions are much larger than the thermal energies, $V(R_l) >> k_B T$, the measured correlation function will be so close to its extreme that only lower bounds for V_l can be given. For example, this situation was met in studies of $VD_{0.781}$ by Pionke et al. [3.80], where the short-range order appears to be determined by a repulsive hard core interaction for nearest- and next-nearest-neighbor pairs. It should also be noted that in this limit of strong correlations a normal Gaussian error distribution of the correlation function leads to a non-Gaussian and highly asymmetric distribution for the interaction parameters.

There is a standing challenge to establish a more realistic Hamiltonian by taking into account the displacements in alloys. While the structural simulation of static displacements can already be performed by steepest descent methods or a straightforward energy minimization, establishing an appropriate Hamiltonian appears to be feasible as mentioned above but it requires one to include also the dynamical properties of the system. [3.16]

3.5 Comments on the Uniqueness of the Models and Methods

The standard MC method based on the principle of detailed balance yields (within manageable statistical errors depending on the computational effort and possible but manageable systematic errors, e.g. "finite size effects" and "scaling") a unique determination of thermal averages, such as, for instance, pair and many-body correlation functions. The following discussion considers the uniqueness or ambiguities if we wish to obtain, conversely, information concerning the structure or underlying interactions from given correlation functions. Recall that by measuring (diffuse) scattering intensities one loses information about the phase factors, and not the structure itself but only pair correlations can be obtained within the kinematic approximation.

Therefore, those possible predictions which are based solely on the pair-correlation functions are of most interest.

The reader is referred to the articles of Welberry et al. [1.20, 3.81], who have illustrated, at least for two-dimensional structures, with impressive pictures, the caveat that (i) because of the restricted information from scattering experiments alone, the determination of the real structure could be impossible in cases where the structure is governed by many-body correlations, while (ii) even subtleties in the pair-correlation function become apparent in the scattering. The obviously non-invertible situation in their examples is characterized by completely vanishing pair correlations at all distances, while other many-body terms are present. It could well be that these extreme examples of possible model structures cannot be represented by a physical Hamiltonian, and simply because of this the inversion is impossible.

Here, with respect to the reverse and inverse MC methods, the discussion is confined – unless explicitly stated – to those states of thermal equilibrium which are homogeneously disordered, allowing only for short-range order of finite range, which decays exponentially in the asymptotic limit, and which can be related to short-range interaction models.

The reverse MC method is supposed to yield a physical and plausible structure but not necessarily the true configuration as examined by the scattering experiment, unless it can be proven that the modeled structures are not only unique in point and pair correlations, but furthermore also in all higher-order correlation functions. Nevertheless, simulated structures have been analyzed with respect to many-body correlations.

The validity of such an approach has been discussed recently by Evans [3.82]. As shown by Henderson [3.83], the pair-correlation function of a liquid determines a pairwise interaction potential to within an irrelevant constant, i.e. the force is uniquely determined. In this case, if a priori only pairwise interactions are relevant, and a solution exists for these, all higher-order correlation functions are uniquely determined. Evans [3.82] suggested that in the more general case all correlations up to the N-body correlation function should be required to determine N-body interactions. It is easy to show for particular although uninteresting cases of perfect order that this latter condition may be necessary, but it is at least not sufficient in all cases. Indeed, it is more interesting to show that the complete information about the N-body correlation could never be sufficient to provide a unique determination of the interaction model.

Theories for liquids [3.84,3.85] have been developed to describe an N-body interaction by effective pair interactions, which describe correctly the direct pair-correlation function and thermodynamic properties via the compressibility equation. The role of three-body potentials in disordered alloys was treated by Taggert and Tahir-Keli [3.86]. They discussed the possibility of representing the three-body interactions by effective pairwise interactions (EPIs) which were composition-dependent in the linear approximation of a generating high-temperature expansion. Their result was that these EPIs do not define the true triplet correlation completely and that the concept of simple composition-dependent EPIs, in particular, becomes invalid as one approaches the transition temperature from above.

It is necessary but straightforward to include higher-order expansions up to arbitrary order in β, for an EPI model from an original many-body interaction model, which leads to identical pair correlations for a homogeneous disordered phase in thermal equilibrium (Fig. 3.1). This has been shown for lattice gas models (Ising models) [3.87] and tested by accurate MC simulations. Hence, for a given pair-correlation function for an equilibrium configuration, it is not possible in any case (no matter how much effort has been spent and how precisely it has been measured) to determine uniquely the relevant interactions if we cannot exclude a priori all many-body terms.‖

‖ In a recent paper [3.88] it has been rediscovered that "effective" interactions as determined from short-range order by the inverse Monte Carlo method are not unique. However, the essential point to note is that the *effective* pair interactions are indeed unique if only a pair-interaction model is used, and they must be different from any other "true or original" pair interactions if they also have to account for many-body effects.

The somewhat unexpected and remarkable result was that not only could the pair correlations be reproduced in rapidly converging approximations, but also the relevant many-body correlations (Fig. 3.2).[§]

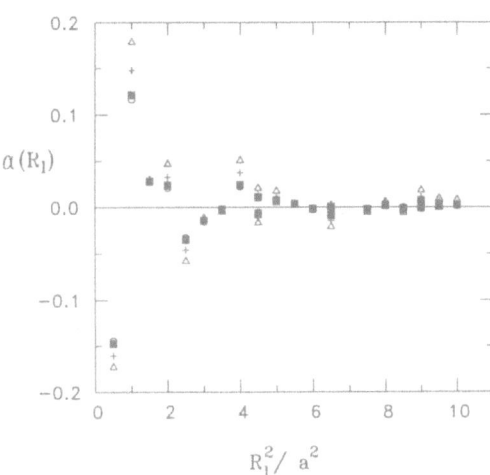

Fig. 3.1. Comparison of the pair correlations originating from a many-body interaction model (*solid squares*) to those of an effective pair-interaction (EPI) model for different levels of approximation, all calculated by Monte Carlo methods. *Triangles*: bare pair interactions only. *Crosses*: concentration-dependent EPI (exact for $T = \infty$). *Circles*, typically coinciding with solid squares: concentration- and temperature-dependent EPI (first expansion in $1/k_B T$; the range of this interaction exceeds that of the original many-body interaction). From [3.87]

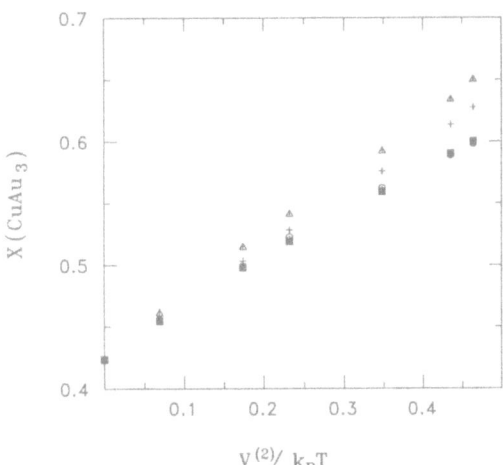

Fig. 3.2. Monte Carlo results for the probabilities of $CuAu_3$ clusters versus temperature for the original many-body interaction (*solid squares*) compared to different approximation levels by effective pair interactions (EPIs) as described in Fig. 3.1. The relevant many-body correlation function described in terms of "$CuAu_3$ clusters" is surprisingly well approximated by the EPI $V(c, T)$ (*circles*). (Reciprocal temperature scale has been normalized by the bare pair interaction $V^{(2)}$.) From [3.87]

[§] The actual many-body interaction model was proposed by Kikuchi and de Fontaine (see [3.89]) to simulate the Cu-Au phase diagram. Here it is applied to a disordered alloy model at 1500 K having the composition $CuAu_3$, where many-body effects are expected to be most significant.

One may note that for this example, the second-order approximation of EPIs also already describes surprisingly well the essential many-body correlations, namely the $CuAu_3$ cluster probabilities, for all temperatures above the ordering transition. The inverse Monte Carlo method, however, would yield the EPIs which correspond to the ultimate approximation in all orders of (inverse) temperature to the original many-body interaction.

This result allows another interesting conclusion. Even if we could know all (including many-body) correlations of a representative configuration, this is not necessarily sufficient to reverse the procedure and determine uniquely the underlying interactions. The case that it never suffices to reverse the procedure would be necessary and sufficient to conclude further that the reverse Monte Carlo method yields the correct and true structure from the pair correlation function.

The question is, whether the pair-correlation function determines uniquely and truly the structure which may have formed according to an arbitrary many-body interaction.

Instead of an Ising-like many-body interaction, a more stringent test should be the choice of interactions with a molecular type of bonding, with comparably small van der Waals interactions beyond the bonded cluster itself. To idealize this situation, let us consider the configurations of non-interacting "lattice animals" consisting of three-atom triangular clusters on a triangular lattice. In this case, the whole pair-correlation function, as easily found, is particularly simple:

(i) the correlation to nearest neighbors is $\alpha_1 = 1/3$,

(ii) all correlations between more distant atoms identically vanish to zero,

(iii) these correlations are independent of the concentration of the clusters and the temperature.

Recall that the recipe of the inverse MC method only permits one to determine uniquely a pairwise interaction model. However, an Ising model with pairwise interactions which could produce such a pair-correlation function, even at only a single composition and temperature, would be very unusual, and would be unable to describe the independence of T and c. A nearest-neighbor EPI leads to a simple monotonic decay (of the modulus) of the correlations with increasing distance, and above T_c its asymptotic form is dominated by the exponential decay as discussed by Ornstein and Zernike [3.90] and Fisher and Burford [3.91]. Therefore, within a purely pairwise model one would need further EPIs which are self-compensating with respect to the pair correlations at all distances other than the nearest-neighbor distance. In other words, we meet the unusual situation where the range of the effective pair interactions exceeds the range of the true pair-correlation function. If one simulates this unusual type of pair correlation in a model using the reverse MC method one can indeed find, with large significance, typical configurations of the triangular "lattice animals" (see Fig. 3.3). Essential information on this higher-order correlation is obviously buried in the extent of the more distant pair correlations even though these are equal to zero. To answer the question of whether the reverse MC method provides exactly the true structure in any case, this example shows that it may hold for the dilute limit but it is not true in the general case. Otherwise, there remains a small but finite number of wrongly bonded atoms, which in practice does not decrease to zero if one includes an increasing range of pair correlations in the simulation.

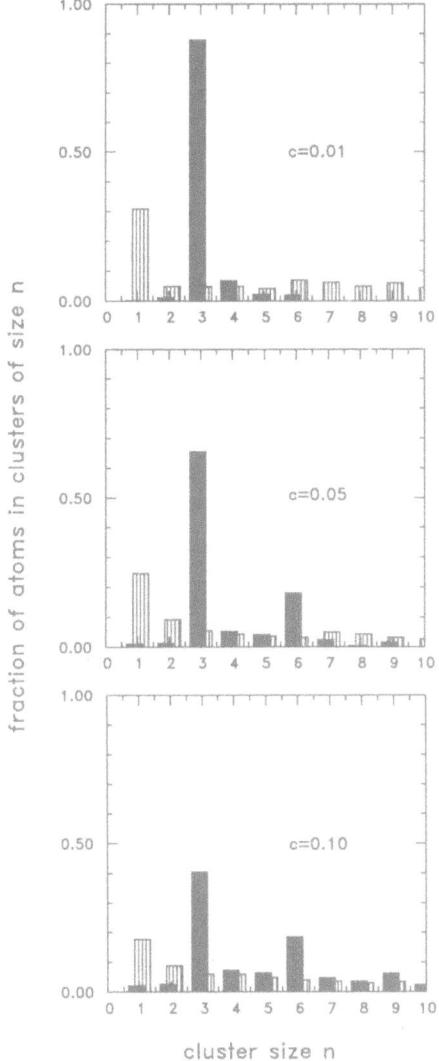

Fig. 3.3. Cluster distributions for reverse MC simulated structures on a triangular lattice with only nearest-neighbor pair correlations $\alpha_1 = 1/3$. The corresponding true, ideal solution of the structure would be a random distribution of triangular clusters, and therefore only clusters of sizes 3, 6, 9, etc. should appear. From *top* to *bottom* the coverage increases from 0.01 to 0.05 and 0.1. The comparison shows that information about multi-site correlations is to be found in the full extent of the pair-correlations (even if these are zero). First (*hatched bars*), only α_1 is simulated, second (*filled bars*), further correlations, $\alpha_l = 0, \quad l > 1, ..., 12$, are included

In using the inverse MC method, one finds slowly converging effective pair interactions, however, and the series of the moduli of the EPIs does not seem to converge. The range of EPIs increases further with the range of simulated pair correlations even though these are zero. While in the case of Ising-like many-body interactions there is a rapidly converging scheme to establish the pair correlations by effective pairwise interactions, there is no obvious converging scheme for such cases of "molecular" bonding and thus no acceptable solution in terms of EPIs.

Of course, despite all successes of the Ising model and its variants, it is not possible to reduce the whole world to it. For instance, the expectation that the Connolly–Williams scheme [3.25] provides a complete and unique mapping of total energies, as calculated from ab initio electronic theories, onto a Hamiltonian of Ising type (see [3.26]) may be a good approximation but is generally unwarranted.

At least, if a scattering experiment reveals a correlation (or partial structure factor) which does not depend as expected for a pairwise model on concentration (or density) and temperature, it need not be too difficult to identify the responsible microstructure and the reverse MC simulation should be able to reveal at least qualitatively the significant structures even in such notorious cases.

Furthermore, it is important to note that the acceptance for the pair exchange becomes rather poor finally (N^{-2}), owing to the hidden and very specific reduction of the configurational entropy, which can be regarded as indicative of the fact that the topology of such a disorder is not formed by an isotropic pairwise bonding. If we had anticipated the particular type of bonding, we could have used it in the reverse Monte Carlo method, finding quantitatively and accurately, in an easy and fast simulation, the desired pair correlations, by moving triangles on the lattice only, rather than by exchanging two atoms. The result would also imply the interaction model. This suggests the need to include new strategies within the reverse and inverse Monte Carlo methods, searching for a solution where the type of Monte Carlo step is altered with respect to a maximized acceptance rate.

Is it possible to determine many-body interactions from pair-correlation functions?

The exact high-temperature expansion scheme as given in [3.87] shows that a given Ising Hamiltonian including many-body interactions, which are thought to be independent of the average composition (concentration or density) and temperature, can be mapped onto an effective pair interaction model, which is dependent on composition and temperature and has a larger range than the original many-body interaction. Hence, the asymptotic effective pair interactions (expanding to all orders) are what is determined by the inverse Monte Carlo method. The approximations should converge rapidly in cases of Ising-type Hamiltonians. The range of the effective interactions should remain small compared with the range of the pair correlation function itself. This can also serve as a good criterion eventually to reject a model consisting solely of effective pairwise interactions.

If a priori one may assume the real interaction (including possible many-body terms) is composition- and temperature-independent, or that it is at least either composition- or temperature-independent, measurements of the pair correlations for different states of temperature and/or composition could yield different effective pair interactions; these could at least verify or falsify suggested many-body interactions. All these conclusions should in principle hold also in the case of liquids (and amorphous materials if the structure is related to a well-defined local minimum of the free energy). Indeed, for the case of liquid krypton, it has already been demonstrated by Barocchi et al. [3.92] that variation of the temperature and the density in scattering experiments provides sensitive information about many-body interactions.

In real alloys the situations are a priori less well defined than in the examples discussed above. Many-body terms are theoretically expected to be small compared to the true pair interactions and therefore difficult to verify from scattering experiments. The recipe, namely to extract the relevant information about the many-body terms from the dependence of the effective pair interactions on temperature and concentration (and density in case of liquids), also implies severe assumptions, for instance that Ising-type models are adequate at all, that there are no other a priori unknown non-local effects on the interaction model, that the bare pair interaction is independent of temperature, that the influence of magnetism is treated in a proper model (see Sect. 4.2.1), etc.

Practical limitations also arise if it is not possible to obtain complete information about the pair-correlation functions. This does not only refer to the statistical accuracy of the scattering data. Interactions can only be determined if these are not too large compared with the thermal energy. Otherwise, only lower bounds can be given. Contrary to normal liquids, pair correlations on lattices are in general anisotropic. The information about the anisotropy can be revealed by studying single crystals. However, in cases where non-equivalent sublattices exist, in interstitial alloys for instance, it may be usual also to have non-equivalent sublattice correlations between equidistant sites and equivalent directions. This might prevent the unambiguous determination of all of the pair correlations (e.g. $Fe_{1-x}O$ or oxygen-vacancy correlations in $YBa_2Cu_3O_{6+x}$).

4. Applications

Various examples are selected here for discussion of how successful approaches to the configurational statistics of alloys can be made on the basis of Ising models, which may still exhibit rather complex behavior despite all the inherent simplifications. Alloy phase diagrams can be calculated by means of Monte Carlo methods or mean-field approximations, namely the Bragg–Williams model or the more sophisticated cluster variation methods, for a given Ising Hamiltonian, and may be compared with the experimentally determined phase diagram. The interactions may be derived from diffuse scattering due to the short-range order in the high-temperature disordered phase. Particular emphasis is given to this approach here. Modern electronic structure calculations have made remarkable progress and lead to predictions of atomic interactions for a number of alloys. Further, there have been many attempts to model phase diagrams by refining the microscopic interaction model with respect to the macroscopic observation of phase boundaries. These approaches aim at the realistic modeling of specific alloy materials, with the hope of useful applications to alloy design in the future. Besides, there are long-standing interests in the generic behavior of order and disorder in alloys. Therefore, universal and critical phenomena, which are thoroughly investigated for most of the standard bulk cases, have been studied by the most simplified interaction models but even these sometimes place extreme demands on the numerical accuracy (see Sect. 3.4).

4.1 Clustering Alloys

The simplest type of phase diagram is found for alloys with clustering tendencies, forming a solid solution on a coherent lattice at high temperatures while at lower temperatures a phase separation into A- and B-rich phases leads to a miscibility gap. Because of this simplicity, such examples are best suited as an introduction to the relevant topics related to our understanding of order and disorder in alloys.

First, to discuss the applicability of the methods, a comparison is made between Monte Carlo results and the Bragg–Williams solution for the phase diagram. With the Monte Carlo method, the phase diagram is easily computed within the grand canonical ensemble with $h = 0$. For a given temperature the system will relax quickly towards $\langle s \rangle = 0$ above T_c and to the miscibility gap below T_c. It may appear more natural to use the canonical ensemble for alloys. Indeed, this would be very suitable for studying the process itself, for instance the kinetics of spinodal decomposition and nucleation and growth. Because of its slower relaxation times this ensemble is less efficient for the determination of equilibrium phase boundaries.

For a discussion of typical discrepancies between the mean-field (MF) solution and Monte Carlo results, various interaction models, which differ in range as well as in the frustration due to competing interactions, were considered in a tutorial by Schweika [3.15]. Figure 4.1 reveals that the MF- and MC-results coincide asymptotically for low T and c. Correlation effects and finite size effects are almost negligible, i.e. long MC runs on small systems yield comparable results to shorter runs on large systems. Larger discrepancies are observed around the critical temperatures T_c. Compared with the MC simulation of a simple fcc nearest-neighbor model, the MF result overestimates T_c by about 25%.

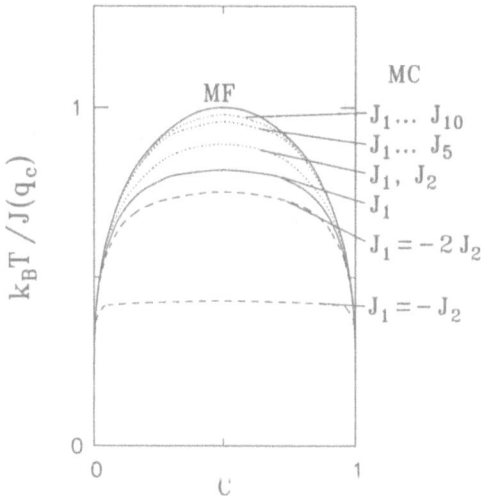

Fig. 4.1. Phase diagrams for alloys of decomposition type. The comparison with the Monte Carlo results (on an fcc lattice) shows that the mean-field solution (*upper curve*) is the envelope of all possible interactions models extending to different ranges, as well as of models with competing interactions

However, considering the critical temperatures, a clear dependence of the MC results is found on the way the total interaction is distributed over neighboring atoms. If the interaction energy is distributed monotonically onto further and further neighbors, the MC and MF results should finally coincide as expected. The opposite is found for oscillating interactions. A competing (negative) next-nearest-neighbor interaction lowers T_c, even if the sum of interaction energies $J(\boldsymbol{q}_c)$ contributing to the critical wave vector $\boldsymbol{q}_c$ is kept constant.

Locating precisely the second-order transition at T_c turns out also to be non-trivial for the Monte Carlo method. Since the correlation length *and* the correlation time diverge at T_c, finite size effects must be analysed and long Monte Carlo runs are required. In contrast to the unique solution based on the Bragg–Williams mean-field approximation for each of various interaction models, the Monte Carlo results will depend on the numerical efforts and may be even more accurate than the results obtained by exact series expansions (e.g. [4.1]).

The cluster variation method (CVM) can provide much better estimates than the simple Bragg–Williams MF solution (which is equivalent to the point-cluster approximation in the CVM). For the nearest-neighbor interaction model the tetrahedron approximation yields $kT_c/|J(q_c)| = 0.83544729$ (with $J(q_c) = 12J_1$), and the next commonly used CVM approximation of two clusters, a tetrahedron and a octahedron, yields $kT_c/|J(q_c)| = 0.83394338$ [1.1, 3.2, 4.2, 4.3]. These results differ by only about 2% from the accurate result obtained by series expansions, $kT_c/|J(q_c)| = 0.81627$ ([4.4]), which is remarkable, since the singular critical properties of the correlations at T_c cannot be treated correctly in such small finite clusters. Nevertheless, the CVM seems to give rather satisfactory results concerning the calculation of phase diagrams. However, the cluster size limits the range of the interaction model which can be appropriately treated, e.g. the tetrahedron CVM may be fitted to the J_1 model only and the tetrahedron–octahedron CVM permits one further to include J_2. In contrast to the Monte Carlo method, it is also cumbersome in the CVM as well as in the series expansion techniques, to account for interactions with more distant neighbors.

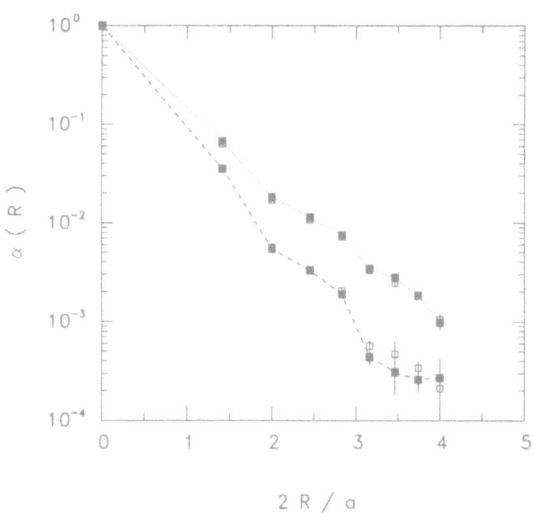

Fig. 4.2. Mean-field theory predicts a scaling behavior which leads to "iso-short-range-states" above a spinodal line. The examples have been calculated by the Monte Carlo method for the nearest-neighbor "ferro" model on the fcc lattice. Comparisons are shown for the short-range order at $T = 27.9$ J/k_B, $c = 0.5$ and at $T = 17.0$ J/k_B, $c = 0.5$ (*filled squares*) with results for $T = 7.0$ J/k_B, $c = 0.048$ and $T = 8.2$ J/k_B, $c = 0.100$ (*open squares, upper* and *lower curves* respectively. Lines are intended to guide the eye)

Figure 4.1 displays only the binodals of the various models, and not the spinodals where the correlation length diverges. The Krivoglaz–Clapp–Moss

(KCM) formula defines also the mean-field spinodal, which is in practice difficult to reach in a Monte Carlo simulation, since as in the real world nucleation
may take place and cannot be avoided in the simulation. (For a closer discussion of interesting, complex kinetic behavior see [4.5].) Furthermore, the
KCM formula implies that there are "iso-short-range-order" curves above the
spinodal. The Monte Carlo calculation shown in Fig. 4.2 gives evidence that
this MF prediction is qualitatively correct, at least to a very good approximation, and that the short-range order has scaling properties even far from
the critical point.

4.1.1 Solubility of Co in Cu

One practical application could be to estimate the interactions from precisely
measured experimental phase diagrams.

For example, the boundaries of solubility of Co in Cu are reproduced
within a 1% deviation by a nearest-neighbor interaction $J_1 = 16.35 \pm
0.12$ meV. Since the deviations with respect to the curvature of the solu-
bility boundary seem to be of a systematic nature, it appears to be justified
to include interactions J_2 to second-nearest neighbors. This yields perfect
agreement with the experimental solubilities [4.6, 4.7], but also demonstrates
the limits of such a method for determining interactions: $J_1 = 21.4$ meV,
$J_2 = -9.8$ meV for CuCo. Models with further interactions would yield ar-
bitrarily many different sets of solutions, since the available experimental
information is already fully exploited.

Recently, Hoshino et al. [4.8] have presented very systematic ab initio
calculations for the interaction energies of impurity pairs of the $3d$, $4sp$, $4d$
and $5sp$ elements in the host metals Cu, Ni, Ag and Pd. (The calculations were
based on the local-density theory and apply the Korringa–Kohn–Rostoker
Green's function method for spherical potentials.) Among this large body
of results – which are particularly valuable for the clustering type of alloy
– precise interaction values $J_1 = 20.1$ meV and $J_2 = -8.2$ meV are also
obtained for the copper-rich Cu–Co alloys, which are nicely consistent with
the experimental solubilities, see Fig. 4.3.

In addition, Hoshino et al. also determined the magnetic interactions
$J_1^{magn} = 19.4$ meV and $J_2^{magn} = -2.6$ meV. One may note that including
interactions to further neighbors does not modify the results mentioned here
in this perturbative approach, in contrast to using the Connolly–Williams
scheme on the total energy results. With all of the interaction parameters
within a double Ising model for occupation and spin variables, a Monte Carlo
simulation yields an almost perfect agreement with accurately measured solubilities [4.7].

Of course, this is a particularly favorable example demonstrating what can
be expected at present from adequate theoretical approaches to the binding
energies in alloys. Furthermore, the large number of results for other dilute

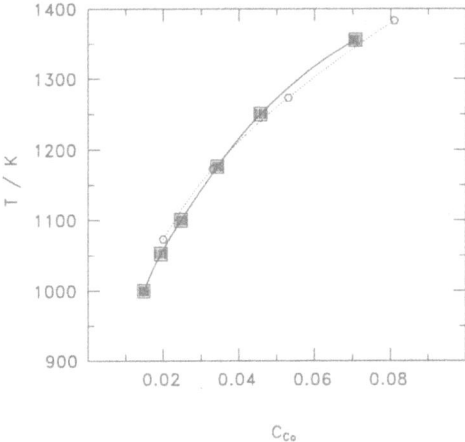

Fig. 4.3. Cu-rich part of the Cu–Co phase diagram. Monte Carlo results (*solid squares*) for the Co solubility in Cu based on the chemical and magnetic interactions predicted by Hoshino et al. agree favorably with the experimental data (*circles*) of Nishizawa and Ishida [4.7]

alloys given by Hoshino et al. agree at least qualitatively with the observed solubilities and trends to specific ordered structures or to phase separation.

4.1.2 Decomposition in the Cu–Ni Alloy System

The Cu–Ni alloys may serve as a first example which demonstrates how effective pair interactions can be obtained from the diffuse scattering by using, in particular, the inverse Monte Carlo method. Particularly noteworthy have been three studies (first by Mozer et al. [4.9], and then by Vrijen, Radelaar and coworkers [4.10] and Wagner et al. [4.11]) of Cu–Ni alloys using null-matrix polycrystalline samples. Their results have been reviewed recently by Moss [1.21], Kostorz [1.17] and Schweika [1.13, 3.71].

These experiments have established a large set of results for various temperatures and different compositions (using different isotopic mixtures). In order to apply the inverse Monte Carlo method one has to be sure that the measured short-range order relates to an equilibrium state. For the $Cu_{0.41}Ni_{0.59}$ alloy investigated by Wagner et al., the interactions $J_1 = (6.1 \pm 0.3)$ meV and $J_2 = (-2.8 \pm 0.3)$ meV have been determined independently from the short-range order [3.71] for three different temperatures, 740 K, 690 K, and 640 K, where the short-range order was very likely conserved during the quench procedure. Figure 4.4 compares the experimental short-range order parameters for $Cu_{0.41}Ni_{0.59}$ at 690 K with the Monte Carlo result based on J_1 and J_2.

The consistency of the results indicates that there are no relevant many-body interactions. In comparison, according to Moss [1.21], there is about a 20% increase in the interaction values obtained from using the $Cu_{0.475}Ni_{0.525}$ data of Mozer et al. The decomposition temperature can be calculated by Monte Carlo methods from these interactions, but this has not been observed experimentally because of the low mobility at these temperatures. A coherent

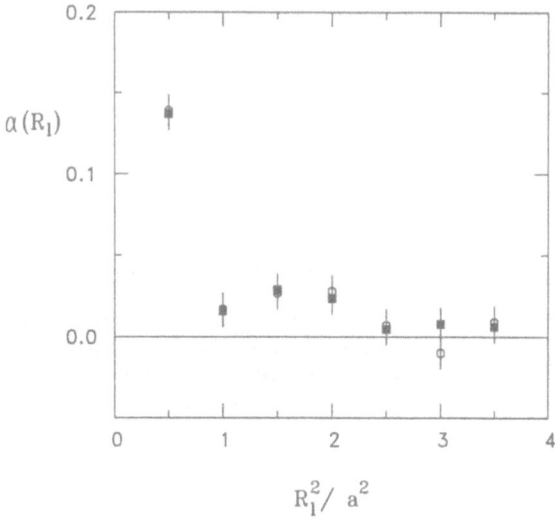

Fig. 4.4. Corresponding pair correlations in $Cu_{0.41}Ni_{0.59}$ at $T = 690\,K$ in terms of short-range order parameters: experiment (Wagner et al., *filled dots*), Monte Carlo result (*open circles*)

miscibility gap is found by MC simulations [3.71] below $T = 495\,K$. This should be asymmetric because of the concentration-dependent interactions.

From the continuous solubility according to the observed phase diagram [4.12], one would deduce a low magnitude of the interactions. Theoretical calculations have been made for the two dilute limits of this alloy [4.8], which are consistent with this expectation. Furthermore, these results support the indicated asymmetry of the effective pair interactions with respect to exchange of Cu by Ni. At the Cu-rich side, the calculations yield $J_1 = 2.5\,meV$ and $J_2 = 2.5\,meV$, while at the Ni-rich side even a weak ordering tendency was found with $J_1 = -5\,meV$ and $J_2 = 0\,meV$.

In addition to the calculation of thermal averages, the Monte Carlo method also enables one to view the characteristic configurations. Figure 4.5 shows the increasing tendency of the alloy to decompose when approaching the critical point. Furthermore, metastable states as well as the evolution of configurations during nucleation and growth or spinodal decomposition may be of interest. Figure 4.5 (*bottom*) also illustrates the temporal evolution of the decomposition of the alloy when it is quenched from high temperatures and annealed below the critical temperature. The time scale in the Monte Carlo simulation is defined by the number of Monte Carlo steps per lattice site. With inclusion of the usually well-known activation energies for diffusion, this artificial time scale can be translated into the real one.

Both examples, Cu–Co and Cu–Ni alloys, have been discussed by neglecting the atomic size effects, distortions and corresponding elastic energies. Such contributions are favorably small in these cases. In general, however, two important consequences have to be taken into account: the appearance of an incoherent miscibility gap occuring at higher temperatures than the coherent phase boundary (see de Fontaine [1.1]) and an important influence on the morphology of the precipitates during the decomposition (see Chen and Khachaturyan [4.13] for mean-field treatments and Fratzl and Penrose [4.14] for Monte Carlo simulations). A detailed review concerning the mechanisms of the decay of unstable and metastable phases such as spinodal decomposition, nucleation and late-state coarsening has been given by Binder [4.5].

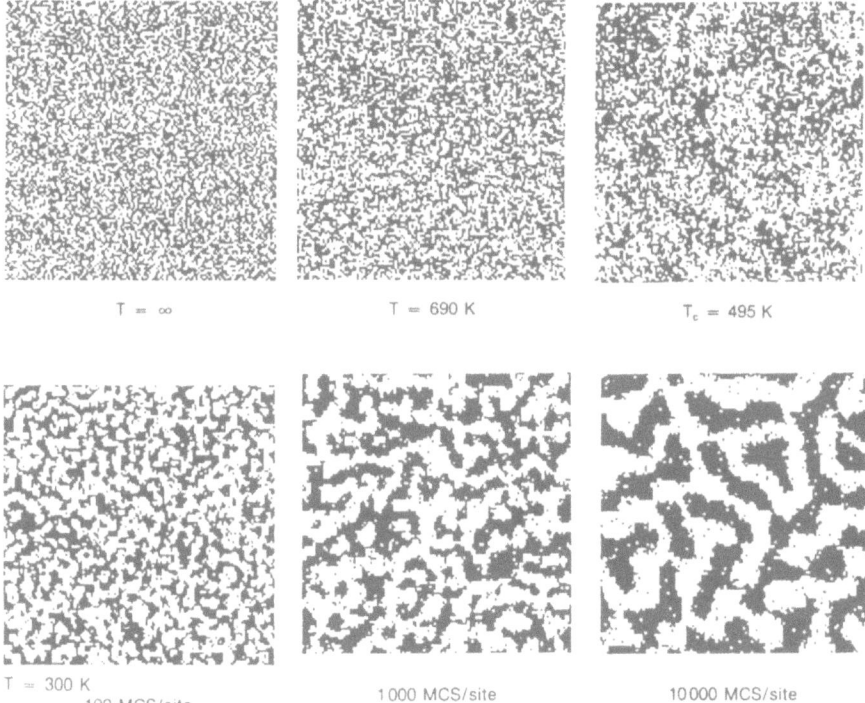

Fig. 4.5. Monte Carlo results for the equilibrium short-range order of CuNi for temperatures $T = \infty$, 690 K and 495 K (*top row*), and for the various stages of decomposition in $Cu_{0.41}Ni_{0.59}$ when the alloy is annealed below T_c at $T = 300$ K (*bottom row*). Configurations are shown after 100, 1000 and 10 000 attempted Monte Carlo steps per lattice site, where exchanges between neighboring atoms were performed. The displayed (100) planes each contain 128×128 atoms and are part of a 3-dimensional lattice. The smallest visible dots represent single atoms. White (black) areas symbolize Cu (Ni) atoms

4.2 Ordering Alloys

For many alloy systems a quantitative description of the configurational statistics needs to incorporate more than pairwise interactions to only nearest and second-nearest neighbors as discussed in the examples above. Furthermore, there may be many other physical ingredients necessary to make the Hamiltonian sufficiently realistic. The following examples will demonstrate how the diffuse scattering of single crystals provides a lot more substantial information which can be accurately analyzed by means of the inverse Monte Carlo method, while perspectives are given on some remaining open questions.

4.2.1 Chemical Order in Ferromagnetic Fe–Al Alloys

This example of a ferromagnetic alloy demonstrates the possible interplay of magnetic and chemical ordering and that the chemical order can even be qualitatively changed by magnetic influences.

The short-range order in body-centered cubic Fe–Al alloys has been investigated in several studies with x-rays, by Semenovskaya et al. [4.15], and with neutron scattering experiments, by Schweika [3.75] and by Pierron-Bohnes et al. [4.16]. The scattering from an $Fe_{0.8}Al_{0.2}$ single crystal at $T = 1013\,K$,

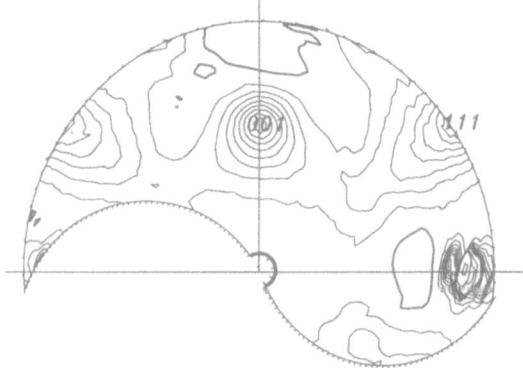

Fig. 4.6. Diffuse elastic neutron scattering intensities at $T = 1013\,K$ of $Fe_{0.8}Al_{0.2}$ in the (110) plane. In the paramagnetic phase the short-range order peaks appear at the ordering wave vector of the neighboring B2 phase, i.e. $h = (0, 0, 1)$ and equivalently at $(1, 1, 1)$

which is above the Curie temperature, is displayed in Fig. 4.6 (see [3.75]). Only elastic intensities are shown, since inelastic scattering due to phonons was separated by time-of-flight analysis. The short-range order peak was found to be at the high-symmetry point $h = (1, 0, 0)$, which shows the alloy's tendency to form the familiar B2 structure of ordered FeAl. With the measured three-dimensional data, a Fourier analysis based on a linear least-squares method yielded short-range order parameters (see Fig. 4.7). Statistical errors are of the order of the symbol size. Negative values of $\alpha(R)$, for instance for nearest neighbors, mean that there is an enhanced probability of unlike Fe–Al pairs compared to the random solution $(\alpha(R > 0)) = 0$.

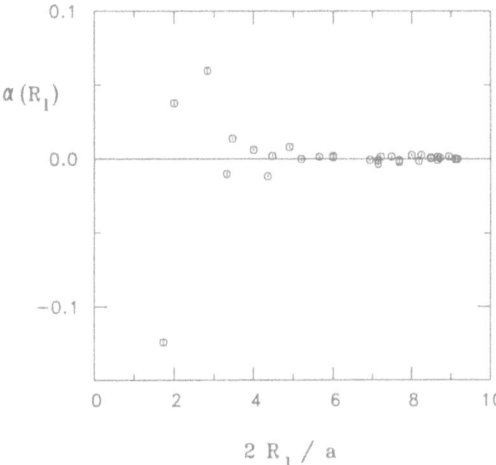

Fig. 4.7. Short-range order parameters of $Fe_{0.8}Al_{0.2}$ at $T = 1013\,K$

In particular, in situ neutron studies at high temperatures [3.75,4.16] also yield quantitative estimates for the effective pair interactions of this alloy (see Fig. 4.8). The inverse MC result shows the vanishing of the interactions beyond the fourth-neighbor shell. The result of the inverse CVM calculation was restricted to the four distinct distances shown which occur in the cluster considered.

In the ferromagnetic state and also slightly above the Curie temperature, the short-range order and thus the effective pair interactions are expected to be influenced by the competing ferromagnetic interactions between the Fe moments. While the chemical interactions are clearly in favor of ordering, i.e. of unlike atomic pairs, the ferromagnetic coupling should favor the neighboring of Fe pairs. Therefore, it is reasonable that the interactions as determined by the inverse MC method [3.75] underestimate slightly the value for the nearest-neighbor interaction. Consequently, the temperature dependence of the effective interactions was investigated [4.16] to obtain an accurate estimate of the pure chemical interactions in the high-temperature limit. The results, which are consistent with those of [3.75], are surprising. It was found that the effective interactions decreased with temperature, although the vanishing magnetic influence should give rise to the reverse trend. The possible origin of the peculiar temperature dependence of the effective interactions in this alloy will be discussed further below.

Attempts have been made to simulate this bcc prototype alloy by Ising models, for instance using the CVM (Contreras-Solorio et al. [4.17]) and MC methods (Dünweg and Binder [4.18]). The agreement achieved with the experimental phase diagram was still fairly satisfactory.

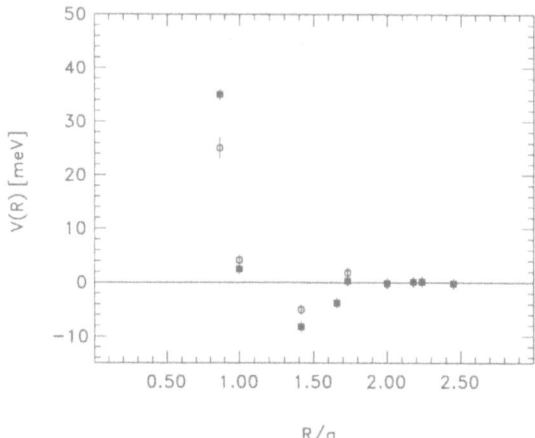

Fig. 4.8. Effective interactions in $Fe_{0.8}Al_{0.2}$. *Filled squares*: inverse MC results determined on the short-range order measured at $T = 1013\,K$ [3.75]; *open squares*: inverse CVM result on the short-range order in a high-T extrapolation [4.16]. The deviation of the effective nearest-neighbor interactions indicates insufficiencies of the simple Ising model. A plausible suggestion would be that the temperature dependence of the effective pair interactions can be related to the gradient of the potential. The direction of change cannot be explained by any different contributions from magnetic interactions between the magnetic moments of iron as would be important below the Curie temperature

Bichara and Inden [4.19] used the effective pair interactions obtained by Schweika [3.75] in a Monte Carlo calculation of the phase diagram, which yielded an improved and qualitatively correct picture of the phase diagram including the two-phase field $A2$–$B2$. The calculated stabilities of FeAl and Fe_3Al were about 20% and 35% respectively too low. Estimates using the interactions as determined from the experimental high-T limit [4.16] gave a similar but slightly less favorable agreement with the experimental phase boundaries. An even more realistic model was applied in the MC calculations of Schmid and Binder [4.20], who obtained comparable results by including the magnetic transitions and by using a classical Heisenberg model for the spins in addition to the experimental interaction parameters [3.75, 4.16]. While these effective (chemical) interactions, displayed in Fig. 4.8 were calculated by neglecting any magnetic terms, different results could be obtained depending on possible assumptions about the unknown magnetic interactions, which may extend further than nearest neighbors. A better understanding of the magnetic interactions in Fe–Al alloys could be achieved by an investigation of the magnetic short-range order in the system using polarized neutrons, as has been done in a pioneering study of Cu_3Mn by Cable et al. [4.21]. Such results could be included in inverse MC simulations to determine simultaneously the magnetic interactions.

However, since the temperature dependence of the interactions determined by Pierron-Bohnes et al. [4.16] is the reverse of what could be expected from a decreasing ferromagnetic influence, other effects should be of concern as well. The question is, for instance – particularly in view of the significant displacements giving rise to the asymmetry of the short-range order peak in Fig. 4.6 – whether the local displacements contribute to the ordering energy in this alloy. Here, on the basis of the data of [3.75], it is found that Al is the "larger" atom. As expected from the positive lattice parameter change with increasing Al, there is a positive displacement field and nearest neighbors are displaced by $\langle u_{0,l=1}^A l \rangle = 0.024\,\text{Å}$. The local situation is in fact rather more complex, since next-nearest neighbors are attracted by Al. However, the contributions of these "size effects" are zero at the ordering wave vectors of highest symmetry (100) and $\frac{1}{2}$(111). Therefore, they do not contribute to the energies of the known ordered structures [1.1, 1.2, 4.15]. A recently proposed, more general, "compressible" Ising model by Chakraborty [3.16] could eventually solve the remaining discrepancies. This model was devised to explain the typical species-dependent displacements in alloys (as observed in recent synchrotron experiments using anomalous scattering variation), and shows that unlike neighbors are displaced towards each other in ordering alloys and the reverse is true for clustering alloys. In addition to a harmonic lattice potential, therefore, a gradient term as a first-order correction for the effective Ising pair interaction was included. The model successfully described the observed displacements. Though it has not been the subject of the first applications [3.16], it is obvious that within the compressible Ising model, the stability of ordered phases will be enhanced, simply because the number of unlike pairs is increased. The model should also be capable of describing the rather typical lattice contraction, as often observed upon ordering, and it should give qualitatively the correct temperature dependence of the short-range order and interactions. In view of the interactions shown in Fig. 4.8, it is conceivable that there may be a rather steep gradient for the nearest interaction V_1. Therefore, revisiting the Fe–Al system using such a more sophisticated Ising model looks promising and could show whether the experimental Fe–Al phase diagram can be quantitatively understood in terms of interactions determined from diffuse scattering experiments including the displacement effects.

One of the reasons for the long-standing interest in Fe–Al alloys is the wide range of stoichiometry of the B2 and DO$_3$ phases [4.22]. More recently, a new long-range ordered structure in this alloy has been observed by Schweika [3.75]. By neutron scattering experiments a rather off-stoichiometric B32 phase was found at temperatures around 350°C in Fe$_{0.8}$Al$_{0.2}$ alloys, see Fig. 4.9. The observation of the B32 type of order was confirmed by an unpublished TEM study of the same alloy by W. Jäger.

The inclusion of explicit ferromagnetic interactions between the Fe atoms leads to comparable configurational energies for the ground states of the B2

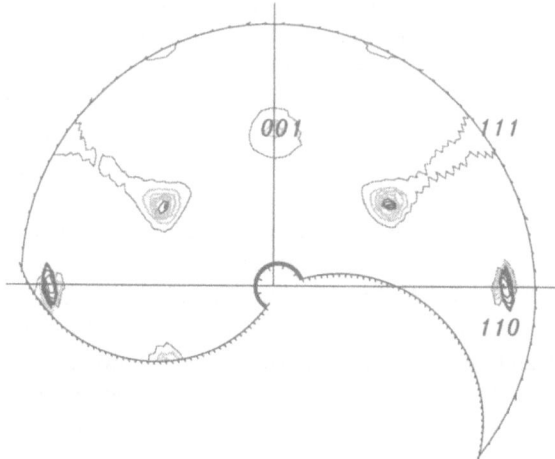

Fig. 4.9. Neutron scattering in the (110) plane of an $Fe_{0.8}Al_{0.2}$ single crystal slowly cooled to 300 K. The onset of a B32 type of order at $(\frac{1}{2}, \frac{1}{2}, \frac{1}{2})$ is observed. It is proposed that this chemical ordering is driven by the ferromagnetism

and the B32 phases at this composition [4.20]. Qualitatively, such an influence of the magnetic interactions on the chemical short-range order towards a B32 type of ordering has also been demonstrated in MC simulations [3.75].

Fultz and Gao [4.23], who investigated by TEM the kinetics of ordering in stoichiometric Fe_3Al, also observed a B32 type of order. This was found, however, to be a metastable transient structure with respect to the finally evolving, stable DO_3 structure. Accompanying MC simulations agreed with the experimental observations. Therefore, they suggested a similar explanation for the B32 observation in the $Fe_{0.8}Al_{0.2}$ alloy. Recent detailed x-ray studies by Becker et al. [4.24] of Fe–Al alloys, with Al concentrations ranging between $0.15 < c_{Al} < 0.2$, also examined the kinetics of ordering and confirmed a stable B32 phase in the iron-rich part of the phase diagram.

In contrast to the work of Fultz and Gao, it is noteworthy that in previous TEM investigations Allen and Cahn [4.25] reported for all of the cases considered by them that the intensity of the ordering wave vector $h = (1, 0, 0)$ increases before that of $h = (\frac{1}{2}, \frac{1}{2}, \frac{1}{2})$. These different kinetic paths become reasonable in view of the annealing temperatures and underscore again the importance of the magnetism in the chemical ordering in this alloy: Fultz and Gao investigated the kinetics of the annealing of their samples below the Curie temperature while Allen and Cahn investigated samples annealed in the paramagnetic state. Hence a realistic modeling of the ordering kinetics in Fe_3Al should include ferromagnetic interactions.

All of the ordered structures B32, DO_3, and B2 are coherent with the bcc lattice and show a superstructure at the high-symmetry points of the bcc lattice; B32 can be characterized by the star – i.e. the group of vectors whose members are equivalent by symmetry – of the ordering wave vector, $h^* = (\frac{1}{2}, \frac{1}{2}, \frac{1}{2})$, DO_3 by $h_1^* = (1, 0, 0)$ and $h_2^* = (\frac{1}{2}, \frac{1}{2}, \frac{1}{2})$, and B2 by $h = (1, 0, 0)$. Figure 4.10 displays the ordered structures which occur in the bcc Fe–Al alloy system.

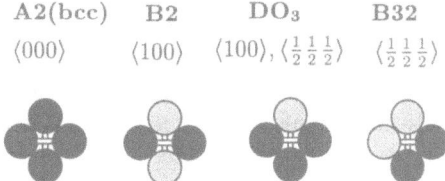

Fig. 4.10. Ordered structures on the body-centered cubic (bcc) lattice and occupations of the four simple cubic sublattices. (The diagram represents a projection of the irregular nearest-neighbor tetrahedron of the bcc structure)

The transitions between the disordered bcc solid solution and the ordered B32 structure and also between the DO_3 and the B32 structure may be continuous according to the Landau–Lifshitz rules:

(i) the space group of the symmetry elements of the ordered structure must be a subgroup of the space group of the parent (less ordered) structure;

(ii) the ordering wave vector must be located at a point of the reciprocal lattice of high symmetry (*Lifshitz points* or *special points*);

(iii) the members of the star of the ordering wave must not add up to a reciprocal lattice vector.

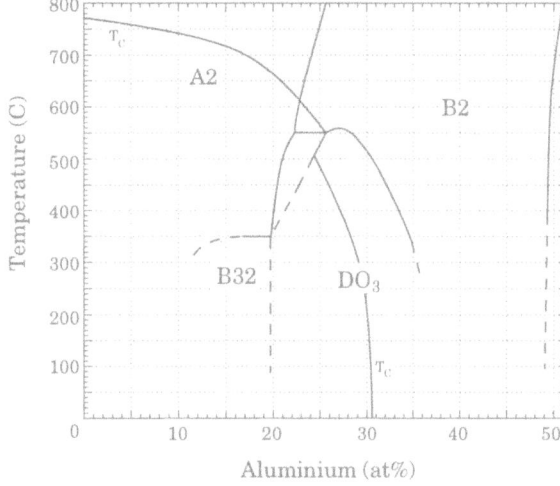

Fig. 4.11. Phase diagram for the iron-rich Fe–Al alloy system. Structures are based on the body-centered lattice; A2 denotes the disordered bcc phase separated into a para- and a ferromagnetic region. In addition to the previously known phases DO_3 (Fe_3Al) and B2 (FeAl), there exists a very off-stoichiometric phase B32 (ideal composition FeAl)

According to the x-ray investigation of Becker et al. [4.24] the transitions from the disordered (ferromagnetic) A2 phase to the ordered (ferromagnetic)

B32 phase are indeed likely to be continuous. If at all, there is only a very weak first-order transition. Figure 4.11 displays the proposed new phase diagram of the iron-rich Fe–Al alloy system. In addition to the new B32 phase, further modifications – which are also in agreement with the work of Oki et al. [4.26] – were proposed concerning the A2(ferro)-DO$_3$(ferro) two-phase field, to be consistent with a more likely continuous transition, also between B32(ferro) and DO$_3$(ferro). The broken lines are tentative since experimental (scattering) data are not available for these transitions.

For sake of completeness, the macroscopic measurements of Köster and Gödecke [4.27] should be mentioned, as these authors have already proposed a new phase or "k-state" in approximately the same region as the B32 phase, which was included in a compilation by Kubashewski [4.28]. Furthermore, Köster and Gödecke [4.27] proposed a subdivision of the B2 phase field on the basis of their measurements, without giving a possible microscopic explanation. Inden and Pepperhoff [4.29] checked this proposal by neutron diffraction, without any noticeable indication of the suggested phase transitions. Therefore, these transitions have not been included in Fig. 4.11.

4.2.2 Short-Range Order and Interactions in the Ni–Cr System

The Ni-rich alloys of the Ni–Cr system are based on the face-centered cubic (fcc) lattice and form solid solutions over a wide range of composition and temperature. However, it has also been observed in this system that short-range order and changes thereof may well have noticeable influences upon macroscopic alloy properties, e.g. the residual resistivity and mechanical properties. Widely used Ni–Cr-based thermocouples may show, for instance, some undesired changes at temperatures below approximately 500°C which depend on the previous thermal history. The ordering tendencies in this alloy system are weak and only a Ni$_2$Cr ordered structure has been observed. The formation of further possible ordered structures, e.g. Ni$_3$Cr, is hindered by the low mobility in the temperature region of interest, at least as far as typical bulk systems are concerned. One may note in this context that it would be of general interest if one could safely predict the equilibrium states of alloys at low temperatures. This could indicate possible long-term changes, under irradiation or in complex artificial or layered structures, where the relevant paths of diffusion could be smaller than in the usual bulk systems.

The disorder in this alloy system has been systematically investigated by diffuse scattering experiments. A rather consistent picture of the effective pair interactions in the disordered phase has been obtained. However, as discussed below, a reliable calculation of the phase diagram has not been achieved and this indicates that a deeper understanding of the ordering mechanism in these alloys is necessary. The experiments have challenged theoretical work to predict the alloy's interaction and its phase stability. However, despite all progress, the theoretical results are far from the desired quantitative level of agreement.

The short-range order in fcc Ni–Cr alloys has been extensively studied by neutrons [1.16, 3.73, 3.76, 4.30] and also by synchrotron radiation [3.77]. The scattering contrast, $\Delta b = b_{Cr} - b_{Ni}$, is favorable for neutrons, but for synchrotron x-rays it can be sufficiently enhanced by anomalous scattering closely below the absorption edges (instead of above, to avoid fluorescence radiation). A similar diffuse scattering was observed in both cases, as shown in Fig. 4.12. Alloys of compositions in the commonly used range $11 - 33$ at % Cr were investigated either in situ or as quenched from not too high a temperature $500°C < T_q < 600°C$ (at lower temperatures the times to reach equilibrium become too long and at higher temperatures the migration of thermal vacancies during the quench may significantly affect the short-range order). The study of Schweika and Haubold [3.73] of the alloy $Ni_{0.89}Cr_{0.11}$ – using the isotope ^{58}Ni to enhance the scattering contrast – revealed that the short-range order maximum is located at $h = (1, \frac{1}{2}, 0)$, which relates this alloy to the Al–Ti ordering family according to the ground state analysis for ordered face-centered cubic alloys by Kanamori and Kakehashi [3.36]. This result agrees with the earlier work of Vintaykin [4.30] on a disordered Ni_2Cr alloy (although a saddle point at $(\frac{2}{3}, \frac{2}{3}, 0)$ was erroneously regarded as the maximum) and was confirmed by all later work on this alloy [1.16, 3.76, 3.77].

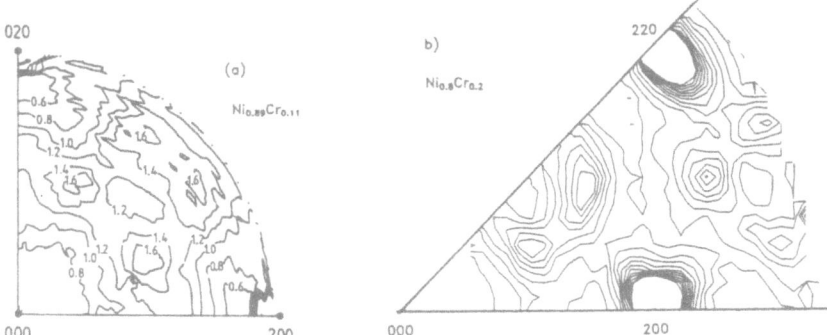

Fig. 4.12. Diffuse scattering of (a) $Ni_{0.89}Cr_{0.11}$ measured with neutrons and (b) raw data for $Ni_{0.8}Cr_{0.2}$ measured with x-rays of energy $E(K_{Cr}) - 20\,eV = 5969\,eV$ at the ORNL beamline X-14 at Brookhaven (from [1.13])

Figure 4.13 summarizes the results for the short-range order as obtained from the various investigations of Ni–Cr alloys. One may note that the extraordinary accuracy of the short-range order as determined by the Fourier methods from such diffuse scattering experiments is certainly superior to that of any other experimental method, such as extended x-ray absorption fine structure, perturbed angular correlation, field ion microscopy.

The scattering data for the alloy $Ni_{0.89}Cr_{0.11}$ was analyzed by the inverse Monte Carlo method, which was further exploited to determine the effective

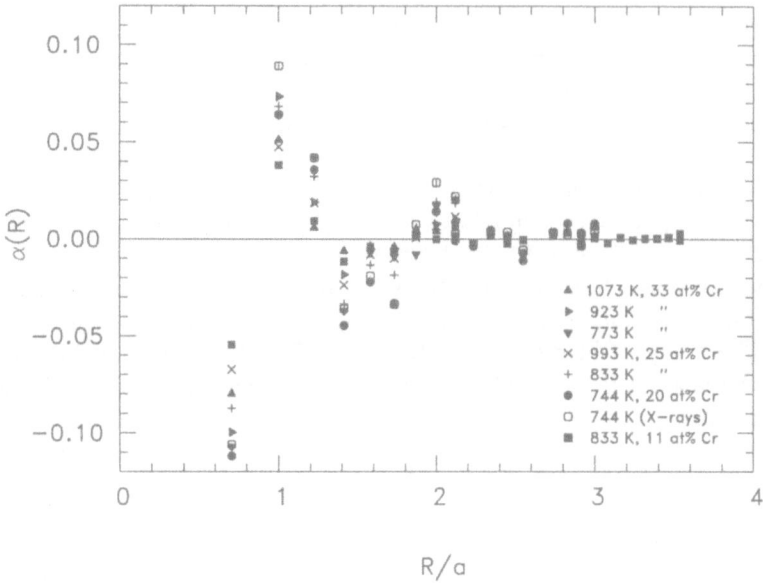

Fig. 4.13. Experimental short-range order parameters for various Ni–Cr alloys (from [1.16, 3.73, 3.76, 3.77, 4.30])

pair interaction out to distant neighbors (see Fig. 4.14). The comparison with the mean-field result using the Krivoglaz–Clapp–Moss formula gave a rather nice agreement, except, as expected, for the largest value of the interaction for nearest neighbors. The γ-expansion method of Tokar, Masanskii and coworkers [3.32–3.34] provides a simple improvement on the mean-field approximation given by the Krivoglaz–Clapp–Moss formula, and also the present case of the alloy $Ni_{0.89}Cr_{0.11}$ was treated by Masanskii et al. [3.34]. Recently, Reinhard and Moss [3.35] reviewed the existing mean-field approaches to determine effective pair interactions – Cowley's theory [4.31], the Krivoglaz–Clapp Moss formula, the improved MF treatment by the γ expansion method of Tokar et al. [3.32–3.34] – and compared them with results obtained by the inverse Monte Carlo method. The applicability of the mean-field approximations will be further discussed below for the example of the bcc CuZn alloys. Alternatively, the inverse CVM (cluster variation method) of Finel [3.5] can be used with comparable accuracy to the inverse Monte Carlo method, as shown for example by the investigation of Caudron et al. [1.16].

For the $Ni_{0.89}Cr_{0.11}$ alloy it was found that the interactions, particularly in the $\langle 110 \rangle$ directions, extend over a few lattice constants, with an oscillating character. Assuming a Fermi surface effect, a Friedel-type model yielded $k_{Fermi} = 1.0\,\text{Å}^{-1}$ and a flattening of the Fermi surface normal to the $\langle 110 \rangle$ directions, see Fig. 4.15, which agrees with the shape of the pure Ni Fermi surface.

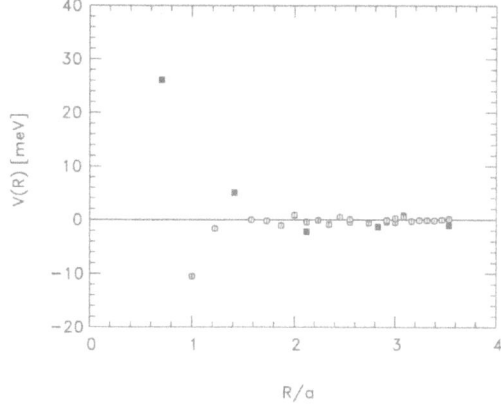

Fig. 4.14. Effective pair interactions in $Ni_{0.89}Cr_{0.11}$ inverse MC results; *solid squares* denote the relatively large values in the $\langle 110 \rangle$ directions (from [3.73])

These findings are consistent with two-dimensional positron annihilation experiments. Theoretical electronic structure calculations (KKR–CPA) by Turchi et al. [4.32] confirmed first that, at the 11 at % Cr composition, the Fermi surface is still reasonably defined and that k_{Fermi} agrees with the suggestion from the Friedel-type model. For higher Cr contents, the Bloch spectral densities became rather smeared. Therefore, any particular Fermi surface effect on the interactions seems to be unlikely for these alloys of higher Cr content. On the other hand, the same calculations would also predict a clustering tendency of the Ni–Cr alloys, in contrast to the observations. These calculations did not include relaxational degrees of freedom and a possible charge transfer between Cr and Ni. Recently in a more sophisticated calculation, Staunton et al. [4.33] showed that indeed the charge transfer is important for understanding the ordering tendency of the alloy. According to their results the alloy should favor ordering, but slightly more ordering, with a concentration wave $h = (\frac{2}{3}, \frac{2}{3}, 0)$ instead of $(1, \frac{1}{2}, 0)$ as observed. This underscores the value of such diffuse scattering studies, since they offer the most detailed microscopic information that can be used to test modern electron theories of alloys.

In order not to leave a wrong impression, it should be pointed out that the Fermi surface effect on the pair interactions is a subtle one. It is, if at all, of only minor importance for the ordering properties of this alloy. The interesting point is that the short-range order (SRO) still may reveal information about the Fermi surface although its influence is weak. Therefore, the SRO seems to be more sensitive in this respect than a particular electron–phonon coupling (Kohn anomaly), as also seen in other examples such as Cu–Pd alloys (see Moss [1.21]).

The comparison with the other investigations of Ni–Cr alloys mentioned above shows that the concentration dependence is apparently weak. Nevertheless, all Monte Carlo simulations with respect to the stability of a *coherently* ordered Ni_2Cr phase yield estimates of the transition temperatures which

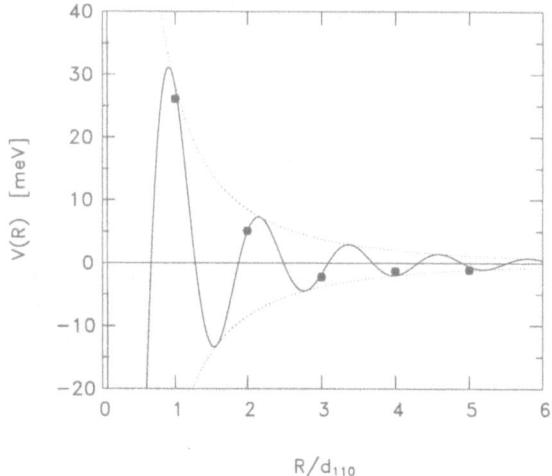

Fig. 4.15. Friedel oscillations in $\langle 110 \rangle$ directions in $Ni_{0.89}Cr_{0.11}$ compared with experimental results. According to the Friedel-type model (*line*) $k_{Fermi} = 10.4\,nm^{-1}$, while the Fermi surface is flattened in the [110] direction. A mean free path of the electrons of $1\,nm^{-1}$ has been used as estimated from the residual resistivity. (From [3.73]) The model is consistent with positron annihilation experiments and calculations of the Bloch spectral density at the Fermi level [4.32]

are about a factor of two lower than the observed transition temperatures of the incoherent phase. The remaining difference indicates the important contribution to the free energy which comes from the relaxational energy accompanying the change from the cubic to the tetragonal structure.

From a methological point of view, the comparative data analysis of a synchrotron experiment on $Ni_{0.8}Cr_{0.2}$ by Schönfeld et al. [3.77] should be mentioned. The results for the effective pair interactions $J(R_l)$ turned out to depend on the data analysis, namely whether the short-range order has been determined either based on the separation methods of Cohen and Georgopoulos or the 3λ method proposed recently by Ice et al [2.6, 2.28]. Furthermore, both results differ significantly from previous neutron data analyzed by linear least-squares methods. Nevertheless, all results agree reasonably well with respect to the short-range order maximum $\alpha(q_c)$, the minimum of the interaction potential $-J(q_c)$ and the calculated ordering transition temperature of Ni_2Cr. Therefore, the systematic errors are likely to arise from the scattering analysis at other scattering vectors and, in particular, from the different ways they separate the comparatively strong thermal diffuse scattering around the Bragg peaks. The ability to separate experimentally this inelastic background using neutrons may emphasize the value of the neutron probe and its permanent need in structural problems despite the new perspectives which are now offered by synchrotrons.

Displacements between the atoms lead to the asymmetry of the diffuse scattering (see Fig. 4.13). It is immediately possible to interpret the pattern qualitatively. If one considers for instance the high-symmetry directions $\langle 100 \rangle$ and $\langle 110 \rangle$, there is an asymmetry across the zone boundary with increasing intensities, $I(q) < I(-q)$ (compare for instance the saddle point intensities $I(\frac{2}{3}, \frac{2}{3}, 0) < I(\frac{4}{3}, \frac{4}{3}, 0)$). The asymmetry changes qualitatively near the Bragg peaks as $q \to 0$ and reverses to $I(-q) < I(q)$ (this is most evident from the low minima at the low-Q side of the Bragg peaks). The latter observation nicely agrees with the macroscopic lattice dilatation with increasing Cr concentration, and the reverse asymmetry at the zone boundary reveals that Cr atoms preferentially *attract* their nearest neighbors which are mostly Ni atoms. A discussion of this particular size effect in terms of Kanzaki force models can be found in [1.13].

The contrast variation using the anomalous or resonant x-ray scattering near the absorption edges can be used in principle to separate the species-dependent displacements for the different atomic neighbors. While the methods of data analysis applied again yield even qualitatively different results in the study by Schönfeld et al. [3.77], the larger reliability was attributed to the 3λ method in this case, which confirms that Cr attracts nearest Ni neighbors, while Cr–Cr as well as Ni–Ni nearest-neighbor pairs are found to be displaced from each other. This means there is *no* simple "size-effect in this alloy and the displacements are mostly due to Coulomb interactions arising from a charge transfer between Ni and Cr rather than due to the repulsive hard cores of the ions. A similar observation was made earlier with the same instrument [2.6, 4.34] in investigations of Ni–Fe (Ice et al. [2.28]) and Fe–Cr (Reinhard et al. [2.43]) alloys. In the ordering alloy Ni–Fe ($J_1 < 0$) unlike nearest neighbors are attracted as in Ni–Cr, and in clustering alloys ($J_1 > 0$), for example Fe–Cr, like nearest neighbors move closer to each other.

An Ising-like Hamiltonian which consistently includes these displacements according to their chemical correlation has recently been proposed by Chakraborty [3.16]. In addition to the harmonic mean lattice potential, a gradient term for the effective pair interactions $J(R)$ needs to be taken into account as a next plausible approximation. It appears likely for the nearest neighbors that the sign of the gradient is equal to that of the interaction itself, i.e. if either like or unlike neighbors are favored, in each case these pairs are displaced towards each other. This would explain the typical species, dependent displacements in alloys. Furthermore, such a Hamiltonian should be used in future inverse Monte Carlo simulations, to determine in addition the gradient term from the species-dependent displacements. The relevance of the displacements to the ordering mechanisms needs to be checked by Monte Carlo simulations of the phase diagram, and a comparison made to the experimental results. One may expect that this more realistic and complete Hamiltonian including all these relaxational degrees of freedom will achieve much better agreement.

4.2.3 The Temperature Dependence of Short-Range Order in Cu–Zn

In the following example we study the temperature dependence of the short-range order in a bcc Cu–Zn alloy in order to check the applicability of the mean-field Krivoglaz–Clapp–Moss formula. The full impact of this formula is that (i) it relates the short-range order to the interactions, (ii) it gives T_c as the denominator vanishes, as well as the spinodal, (iii) it predicts iso-intensity curves of short-range order as a function of temperature and composition and (iv) it contains the Curie–Weiss law for the susceptibility. Recall that $\alpha_l = (\langle s_m s_{m+l} \rangle - \langle s_m \rangle^2)/(1 - \langle s_m \rangle^2)$ and $k_B T \chi(\boldsymbol{q}) = \alpha(\boldsymbol{q})$.

Of course, close to the critical temperature the mean-field approximation fails and the power-law divergence of the short-range order scattering given by

$$\alpha(\boldsymbol{q}_c) \propto (1 - \frac{T_c}{T})^{-\gamma} \tag{4.1}$$

is described for CuZn by the critical exponent of the three-dimensional Ising universality class, $\gamma = 1.24$ [3.91, 4.35]. According to the famous universality hypothesis [4.35], the universality class of a system with short-range interactions is determined by the dimensionalities of the space and of the order parameter, i.e. the symmetry of the ground state.

However, the question to examine is whether the mean-field approximation – which implies that γ and the (non-universal) critical amplitude equal one – holds at higher temperatures or at least in the limit of infinite temperatures, a point of view which is met quite often. According to the Ginzburg criterion [4.36] the mean-field regime terminates where the correlations exceed the range of the interactions. From theoretical work analyzing high temperature series expansions, by Arrott [4.37], and Padé approximants, by Fähnle and Souletie [4.38], it has indeed already been shown that for various models with nearest-neighbor interactions the asymptotic critical exponents should provide a much better description than the mean-field exponents even far from T_c.

There are not too many real alloys exhibiting continuous phase transitions. Historically, it may be worth mentioning that the alloy β-brass served for the first experimental determinations of the critical exponent γ. The first experiments on β-brass by Walker and Keating [2.8] – though suffering from zinc evaporation – indicated that it was not the mean-field exponent $\gamma = 1$ but $\gamma \approx 1.24$ which seemed to be more consistent with the data. Later, β-brass served for the first precise experimental determination, by Dietrich and Als-Nielsen [4.39], of the critical exponents $\gamma = 1.24$ and $\nu = 0.63$ for the three-dimensional Ising universality class.

In neutron diffuse scattering experiments Lamers and Schweika [4.40] investigated the temperature dependence in particular of the short-range order peak near to the melting temperature. The scattering at $T = 808$ K as measured in the (100) plane of the reciprocal lattice is shown in Fig. 4.16.

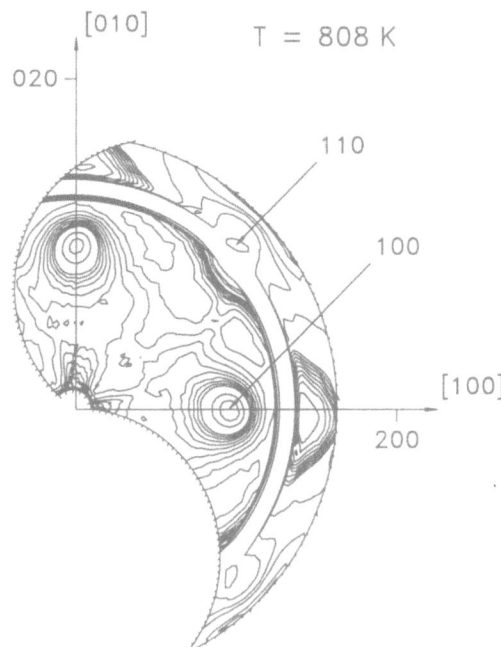

T = 808 K

[010]

020

110

100

[100]

200

Fig. 4.16. Diffuse scattering of bcc CuZn at $T = 808$ K in (100) plane. Short-range order maxima appear at $h = (1, 0, 0)$ and $(0, 1, 0)$. (In addition, there are ridges of diffuse scattering due to the very low acoustic transverse phonon branches in [110] direction running from 020 to 200. There is also a Debye–Scherrer ring due to the Mo sample containment)

Figure 4.17 summarizes all further experimental results for the peak heights (in Laue units) and the peak widths versus temperature. One may note that the data were plotted on an absolute scale in Laue units and extrapolate correctly to the value 1 for a completely disordered alloy at infinite temperatures ($t = 1$). Close to T_c, careful consideration of the resolution was required. The data were obtained on the IN-8 instrument at ILL, Grenoble, and at the DNS spectrometer in Jülich. It was found that a single power law with $\gamma = 1.24$ describes the diffuse scattering in the whole temperature range of the disordered solid β-brass alloy up to the liquidus. Additionally, MC simulations using a nearest-neighbor interaction model for this bcc alloy confirmed that these results persist even for the hypothetical short-range order (above the melting point) in the high-temperature limit.

The correlation length, as obtained from the peak width of the short-range order peak (as the half width at half maximum of the peak intensity), should obey a power law in the asymptotic region:

$$\xi \propto \left(1 - \frac{T_c}{T}\right)^{-\nu} ; \tag{4.2}$$

ξ does not vanish in the high-temperature limit but extrapolates to the range of the interaction. Note that for a nearest-neighbor interaction, the correlation length shown in Fig. 4.17 was normalized by its value at infinite temperature π/a, where $\alpha(\mathbf{R}_1) \cos \mathbf{Q} \cdot \mathbf{R}_1$, due to the nearest-neighbor correlation, is the leading modulation of the Laue (self-)term. Again the results

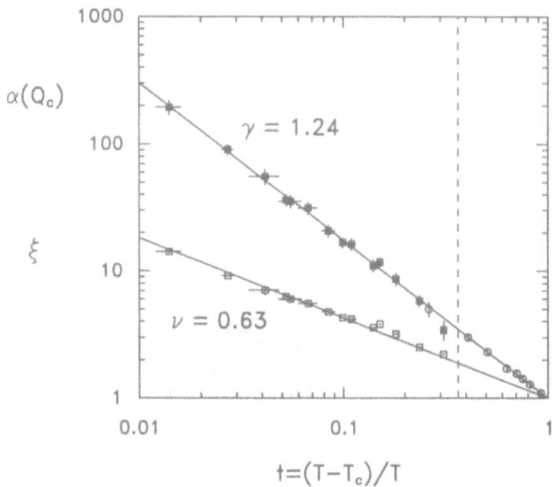

Fig. 4.17. Temperature dependence of the measured short-range order scattering at $h = (1,0,0)$ and the normalized correlation length ξ above T_c in bcc CuZn (β-brass). The solid lines correspond to power laws $\alpha(q_c) = t^{-\gamma}$ and $\xi = t^{-\nu}$ with the three-dimensional Ising exponents $\gamma = 1.24$ and $\nu = 0.63$; the *dashed line* indicates the melting point of the alloy. MC results (*open circles*) represent the hypothetical short-range order above the liquidus in the high-T limit. Note that the slope is not -1 for infinite temperatures as predicted by mean-field theory (Curie–Weiss law)

are consistent overall with the asymptotic critical exponent $\nu = 0.63$ of the three-dimensional Ising universality class.

Note that ab initio calculations of the interactions in β-brass by Turchi et al. [4.41] agree with such a short-range interaction model.

Apparently the scaling corrections are small in the present case, as also found in other instances [4.37, 4.38]; this has led to the suggestion of a "generalized Curie–Weiss-law", replacing the MF exponent $\gamma = 1$ by that of the relevant universality class. For instance, examples of the CuAu and Cu₃Au types of alloy are not expected to exhibit the usual exponents of the three-dimensional Ising universality class, as discussed in [3.15].

Since one may approximate $(1 - T_c/T)^{-\gamma}$ by $(1 - \gamma T_c/T)^{-1}$, one readily sees why the Krivoglaz–Clapp–Moss formula still provides a fairly good description of the temperature dependence of the short-range order and, further, why there is an typically an overestimation of transition temperatures $T_c^{MF} \approx \gamma T_c$ and of $V(q_c)^{MF} \approx \gamma V(q_c)$.

Finally, the non-classical behavior for a square lattice model and the modifications due to increasing interaction range will be discussed. The Monte Carlo data for the high-temperature susceptibility are displayed in Fig. 4.18. Again, the data do not show significant deviations from a power law with the exact critical exponent $\gamma = 7/4$ for the two-dimensional Ising universality class. On the other hand, if the interaction energies are distributed to further

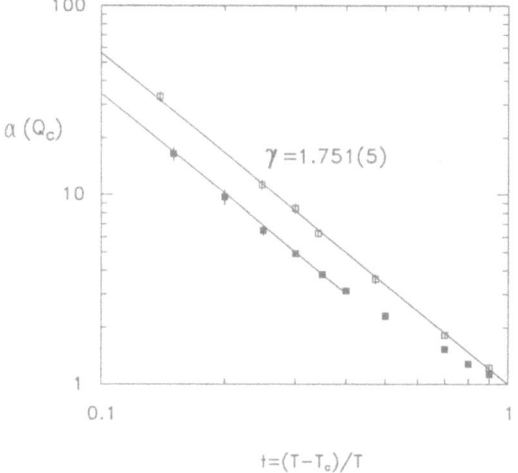

$$\gamma = 1.751(5)$$

$$t = (T - T_c)/T$$

Fig. 4.18. Monte Carlo results for the high-temperature susceptibility of the square lattice. For a nearest-neighbor ferromagnetic interaction model (*open circles*) the exact critical exponent $\gamma = 7/4$ well approximates the entire temperature region, while for a medium ranged interaction model (*solid squares*), extending to the sixth neighbor shell $(J_1=J_2=J_3=2J_4=J_5=J_6)$, the asymptotic Ising region is enclosed by a crossover region for $t \geq 0.4$. Note that at $T = \infty$ the slope is still different from mean-field predictions

neighbors one notices again an Ising regime for $T \rightarrow T_c$. However, systematic deviations from $\gamma = 1.24$ occur for the crossover region for $t \geq 0.4$, while the slope at infinite T ($t = 1$) has still not reached the mean-field value. In this particular example the product of interaction and coordination number was kept constant for the nearest six neighbor shells.

Recently, the crossover from the asymptotic critical region to the classical mean-field limit has been analysed theoretically [4.42, 4.43]. Their asymptotic crossover model is based upon a renormalization group method to first order in the perturbation $\epsilon = 4 - d$, where d denotes the dimensionality of the system. Scaling within this crossover model is obtained by rescaling the susceptibility and the reduced temperature by a Ginzburg number parameter, which is, however, not obviously related to the range of the effective interactions. A variation of the effective interaction range can be achieved in real materials, for instance in binary solutions of various polymer mixtures prepared with different degrees of polymerization. Therefore, such suitable model systems have been investigated systematically by light scattering and neutron small-angle scattering experiments; all results can be nicely scaled with the proposed theoretical scaling function, as shown by Meier et al. [4.44]. For the examples discussed above, i.e. the CuZn alloy and the nearest-neighbor interaction model for the square lattice, which show only very small deviations from the asymptotic critical behavior even for infinite T, the Ginzburg number would be expected to be of the order of 10^2 to 10^3, which is large not only compared with those of the polymer blends but also with those of simple liquids (for comparison a recent review by Sengers [4.45] is recommended).

4.3 Interstitial Alloys

From a conceptual point of view the following two examples of interstitial alloys represent pseudo-binary systems, in which a lattice gas of vacancies and only one alloy component determines the configurational disorder of the system. Therefore, both cases can be treated like the true binaries. It is merely necessary to include the displacements exerted by the defects on the atoms of the fully occupied sublattices. In diffuse scattering patterns, contributions from these displacement fields appear as typical "fingerprints" such as charged defects in ionic systems like $Fe_{1-x}O$ which polarize the near neighborhood. One may note that because of this the short-range order among hydrogen in metal hydrides would become visible in x-ray scattering experiments. However, in the example of VD_x below, we still prefer to use neutrons and the heavy hydrogen isotope D for a direct observation of the short-range order.

4.3.1 The Defect Structure of Ferrous Oxide $Fe_{1-x}O$

What can be learned about the atomistic structure and local order that is relevant to practical interests in metal oxides? One motivation could be that, although the black wüstite $Fe_{1-x}O$ is not the usual reddish-brown rust, an understanding of its microstructure at the atomic level and of the type and behaviour of the defects provides an insight also into processes of oxidation and corrosion, e.g. morphology changes such as that from metals with smooth surfaces to possibly porous ceramics. Here we meet a situation, where, as already known, cations and cation vacancies constitute the mobile defects. Upon oxidation the iron atoms have to migrate from the metal through the oxide to the surface, where oxygen deposits from the gaseous phase. A more subtle problem is that in those cases where the fraction of defects is not small, as is typical and unavoidable in $Fe_{1-x}O$, correlations among the defects may have an important influence on the ionic transport properties. In fact, there is a very unusual and only minor change of the cation diffusion coefficient with increasing deviation from stoichiometry, i.e. with increasing number of cation vacancies. The present study reveals the strong correlations and interactions among the charged defects in this ionic material which are responsible and provides an explanation for its unusual transport properties.

In summary, in situ scattering experiments of the defect structure in non-stoichiometric ferrous oxide $Fe_{1-x}O$ will be discussed. First, the concentration dependence of the scattering from a model calculation based on a typical proposed defect cluster motif, including its long-range displacement fields, will be described qualitatively. Secondly, the equilibrium distribution of the defects by Fourier analysis and computer modeling of the short-range order will be analysed quantitatively. The Fe–O phase diagram (see Fig. 4.19) exhibits a stable high-temperature phase, wüstite, which is well known for its large and unavoidable non-stoichiometry (of at least $x > 0.05$ to more than $x > 0.15$).

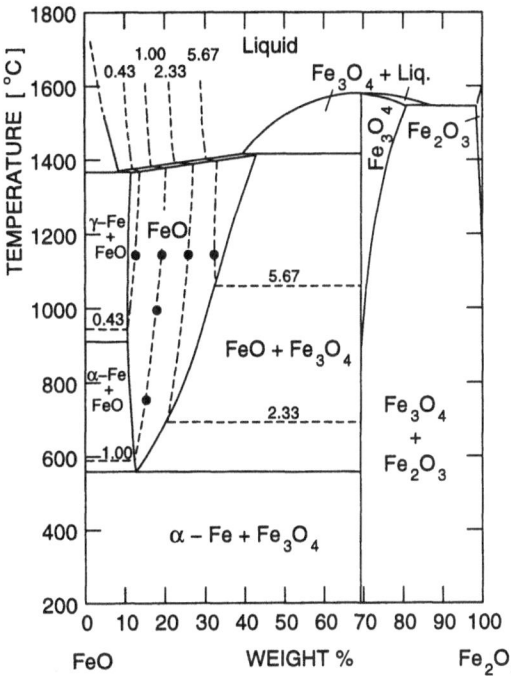

Fig. 4.19. Phase diagram of iron–oxygen after Muan and Osborn [4.47]. Investigated states are indicated (*dots*) in the wüstite phase field. The *dashed line* denotes the ratio of CO_2/CO pressures, which can be used to achieve equilibrium conditions (from [4.46]). Note that FeO (wüstite) does not exist at ideal stoichiometry and has a broad range of stability at high temperatures

There exists a vast literature treating experimentally and theoretically the possible defect structures related to this non-stoichiometry (for detailed references see [4.46, 4.48]). According to powder Bragg diffraction data (see Cheetham et al. [4.49] Radler et al. [4.50]), the fcc oxygen sublattice is fully occupied, while vacancy defects occur in the fcc Fe sublattice and, further, a substantial fraction of Fe ions are tetrahedrally coordinated interstitials. Because of these types of defect and the similarity of the crystal structure of $Fe_{1-x}O$ (NaCl structure) to the spinel structure of the neighboring Fe_3O_4 phase, attempts have been made to explain the possible defect aggregates as precursors of the spinel structure. Most of the proposed defect models are based on the so-called 4:1 cluster, which denotes four cation vacancies surrounding one Fe interstitial. Such a cluster is expected to have a net charge of -5, since the interstitial is supposed to be trivalent and the vacancies together carry eight missing positive charges. Charge compensation can be achieved in such models by either some additional trivalent substitutional Fe ions in the cluster's neighborhood or by a further agglomeration of 4:1

cluster units into larger defects which, by the sharing of common vacancies, improves the charge balance.

Despite such elaborate concepts existing in the literature, the proposed defect structures display an astonishing variety. This "zoo" of defects resulted mostly from speculations based on the mean sublattice occupations as determined from Bragg diffraction, and from simplifying models for the diffuse scattering. In addition, the defects are rather mobile at high temperatures and the equilibrium structures cannot be conserved reliably during a quench to room temperature. Theoretical calculations by Catlow and Stoneham [4.51], Grimes et al. [4.52] and Press and Ellis [4.53], predicted a high binding energy of $\approx 2\,\mathrm{eV}$ for defect clusters of the 4:1 type and variants of it. The large stability of these defect clusters seems to be at variance with the high cation mobility, which led to questionable diffusion models (Stoneham [4.54]). Furthermore, one would like to understand the nonlinearity of the diffusion coefficient with increasing defect concentration, as observed by Chen and Peterson [4.55].

A better insight into the structural defect properties can be obtained by in situ investigations of the diffuse scattering using single crystals. Neutrons are especially favorable compared with x-rays, as used previously by Cohen and coworkers [4.56–4.58], in view of the high background of phonon scattering and the possibility of separating this background by energy analysis of the neutron scattering. In a recent neutron investigation of the diffuse scattering by Schweika et al. [4.46], flowing gas mixtures of CO and CO_2 were utilized to achieve the necessary oxygen partial pressure for varying the non-stoichiometry of an $Fe_{1-x}O$ single crystal at high temperatures. At $T = 1150°C$ the concentration dependence of the diffuse scattering was studied in the range $0.054 < x < 0.131$, which means that the single crystal had to grow and shrink in the atmospheres by about 10% in volume.

The observed diffuse scattering in the (100) reciprocal plane is shown in Fig. 4.20. The bcc Brillouin zones of the fcc lattice are indicated in the figure to facilitate a qualitative understanding of the scattering. Recall that even functions are related to the short-range order and odd functions to the (first-order) displacement terms.

Neglecting first all anti-symmetries and considering only that the diffuse scattering is strong around the 020 and 002 zone centers but weak around 000 and 022, it is readily apparent that the modulation of the Laue intensity does not follow the translational symmetry properties of the Brillouin zones. The doubling of the periodicities clearly indicates that tetrahedral interstitials are present and that there is a strong Fourier component for the nearest-neighbor correlation between the interstitial and the surrounding cations or vacancies.

A second indication of the presence of interstitials can be deduced from the particular asymmetry of the diffuse patterns which is observed not only across the mirror planes perpendicular (and parallel) to the scattering vector Q. There is an additional asymmetry across the diagonals along the (110) directions, which are zone boundaries for the simple cubic interstitial lattice and relevant nodes for the observed displacement scattering terms.

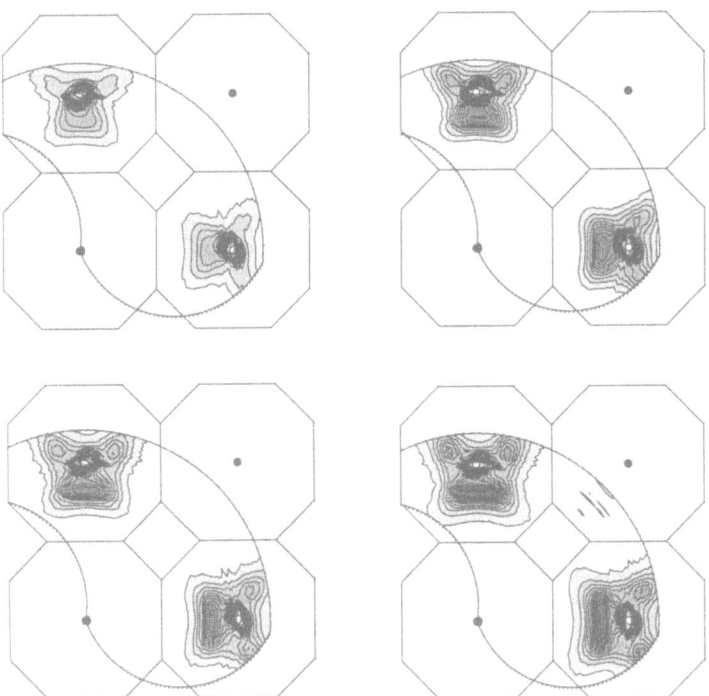

Fig. 4.20. Observed diffuse elastic scattering intensities of $Fe_{1-x}O$ at $T = 1150°C$ in the (100) plane: (*top left*) $x = 0.054$, (*top right*) $x = 0.079$, (*bottom left*) $x = 0.104$, (*bottom right*) $x = 0.131$. (Additional Bragg scattering occurs at the zone centers.) Contours are given in steps of 0.1 b/sr. The bcc Brillouin zones of the fcc cation sublattice are indicated in the figure. Vacancy–interstitial correlations (doubling of the reciprocal cell) and displacement fields (anti-symmetries) are required to explain the missing symmetry properties of the Brillouin zones for the diffuse scattering. (From [4.46])

With increasing non-stoichiometry this qualitative appearance is preserved while distinct diffuse peak intensities evolve around the 002 and 020 Bragg peaks. Since there is a variation of the wave-vector-dependent scattering with concentration, any model calculation within the single defect approximation must be insufficient.

Various types of model calculations as found in the literature have been tried without achieving any satisfactory result. The finally successful model [4.46] was based on a few plausible assumptions only. It was assumed that negatively charged cation vacancies lead to a displacement field which is characterized by repelling the nearest oxygen ions and attracting the cations in the next-nearest-neighbor shell. Within a Kanzaki-type model the elastic properties of the lattice were taken into account by a shell model similar to a model proposed by Kugel et al. [4.59] based on their phonon measurements of FeO. Two Kanzaki forces for the cation vacancies were applied, consistent with the known variation of lattice parameter with non-stoichiometry, by which the entire long-range displacement fields can be taken into account in linear response. The vacancy displacement fields were linearly superimposed for a 4:1 defect cluster.

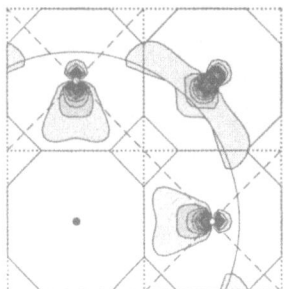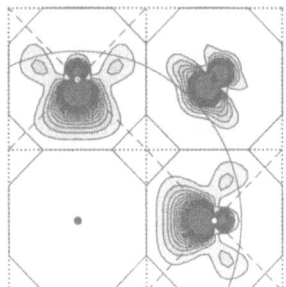

Fig. 4.21. Kanzaki model calculations within the usual single defect approximation for random cation vacancies (*left*) and random 4:1 defect clusters (*right*). The symmetry properties of the latter model are close to the experimental observation, however, the steep increase near the reciprocal lattice vectors (Huang scattering) is at variance with experiment. Intensities are scaled to the non-stoichiometry $x = 0.08$. (Circle segment denotes the experimentally investigated region around the origin. From [4.46])

As a comparison of Fig. 4.20 and Fig. 4.21 reveals, the essential features of the observed diffuse scattering can apparently be related to such 4:1 defect clusters. The most important deficiency of the model is, however, that the calculated Huang scattering is too strong and simply scales with the non-stoichiometry, contrary to observation. The Huang scattering is constrained by the defect strength, which is related to the lattice parameter change per defect, and essentially it does not depend upon the particular choice of the Kanzaki forces. This shows that the approximation devised for the dilute limit does not hold here, and correlations among the defects need to be taken into account. Therefore, it cannot be sufficient to consider only the vacancies and interstitials as defects. All occupational fluctuations need to be taken into account in a consistent and equivalent manner.

To account for the screening of the charged clusters, a "decorated" cluster model was finally devised, in which a 4:1 cluster is definitely surrounded by two shells of iron instead of having a mean probability of also finding vacancy neighbors in these sites. Not only do these fluctuations from the mean site occupancies enter into the short-range order scattering but also, with respect to the mean lattice, the surrounding iron ions exert Kanzaki forces of reversed sign to those of the vacancies. Precisely, the forces are proportional to the occupational fluctuations, $-c^{Fe}$ for the vacancies and $1 - c^{Fe}$ for the cations.

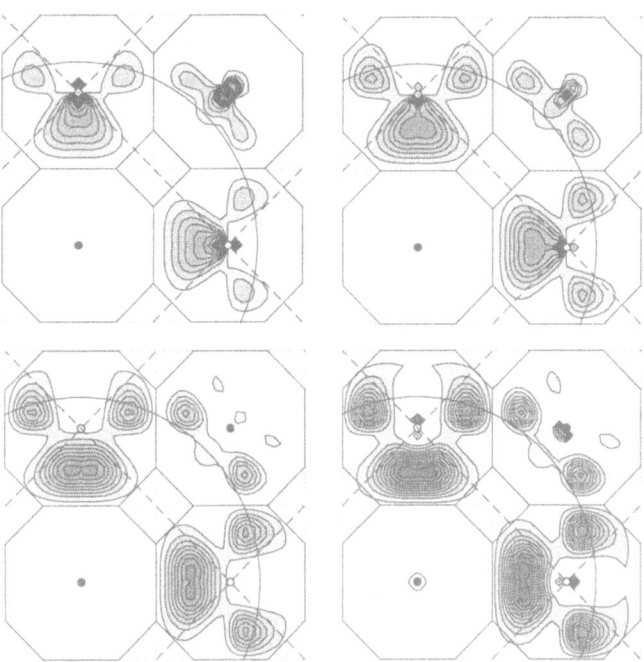

Fig. 4.22. Kanzaki model calculations (with respect to the mean lattice properties) assuming that 4:1 defect clusters are preferentially surrounded by cations for charge compensation. Note that the diffuse intensities for the (100) plane are based on same model for all non-stoichiometries $x = 0.054$ (*top left*), $x = 0.08$ (*top right*), $x = 0.104$ (*bottom left*), and $x = 0.13$ (*bottom right*). (While this very simplified model describes the experimental findings at least qualitatively in many details, the precise wave vector could be reproduced only by the true, more complex defect structure. From [4.46])

As shown in Fig. 4.22, this model can explain the observation which, initially, seemed surprising: the Huang scattering vanishes with increasing defect content. This can be understood as resulting from a (mutual) screening of the charged defect clusters and of their long-range displacement fields. Furthermore, this model also accounts rather well for the characteristic wave-vector-dependent changes of the scattering [4.46]. Actually, the precise wave-vector of the diffuse peak is not correctly predicted from such a crude, simplified model of the defect structure. Recall that only two force parameters, which were chosen to be consistent with the lattice parameter change, give this simplest model to explain the characteristic displacement fields of the charged defects, where effectively negatively charged vacancies repel the (nearest) oxygen ions and attract the (second nearest) iron ions.

A quantitative analysis of the structure was also possible. For one composition of $x = 0.079$ a three-dimensional and sufficiently extensive database was available to perform a Fourier analysis of the short-range order and linear displacement terms. The short-range order parameters displayed in Fig. 4.23 reveal in particular a strong correlation between nearest vacancy and interstitial neighbors. (The negative value of the next-nearest-neighbor correlation between vacancies and interstitials is at variance with the much too simplified screening picture of the model discussed above.)

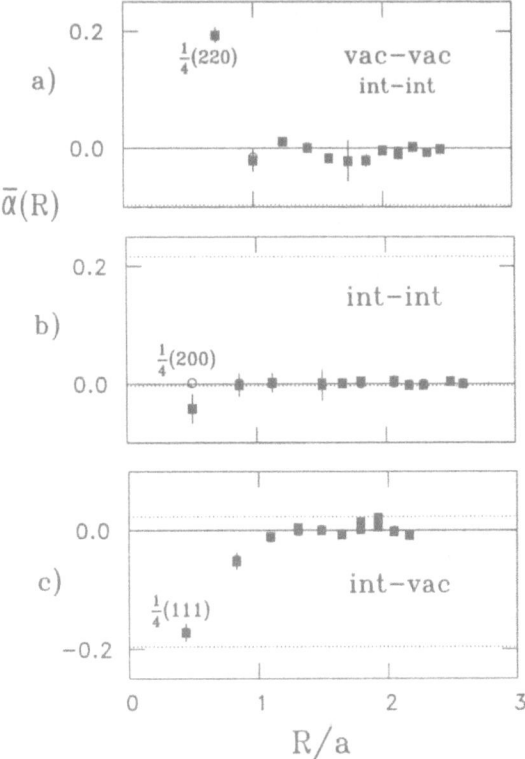

Fig. 4.23. Short-range order coefficients $\bar{a}_l$ for $Fe_{0.92}O$ at $T = 1150°C$: result of the Fourier analysis (*filled squares*) and simulation (*open circles* only one can be seen separately and all others coincide with filled squares). Dotted lines denote lower and upper bounds. (a) Correlations on the fcc sublattices; these are dominated by correlations among vacancies, while a minor contribution comes from correlations among interstitials at the same distances. For example, vacancy neighbors are favored. (b) The part of the correlations on the simple cubic sublattice which is purely between interstitials, showing only random correlations. (c) Interstitial–vacancy correlations, e.g. a very strong attraction between interstitial and vacancy neighbors is found. From [4.46]

The defect correlations were simulated in a computer model of the $Fe_{0.92}O$ structure. All measured short-range order parameters were used for this purpose, applying a reverse MC algorithm and a large model containing 153 600 lattice sites. The resulting structure, e.g. Fig. 4.24, may be analyzed in various ways or just viewed by eye. In view of a possible formation of interstitial–vacancy clusters, which is a non-trivial *many-body* correlation, one would wish to analyze the connectivity of interstitial and vacancy defects. The algorithm for this purpose has been borrowed from percolation-type problems and is similar to that of Hoshen and Kopelman [4.60]. All defect aggregates which

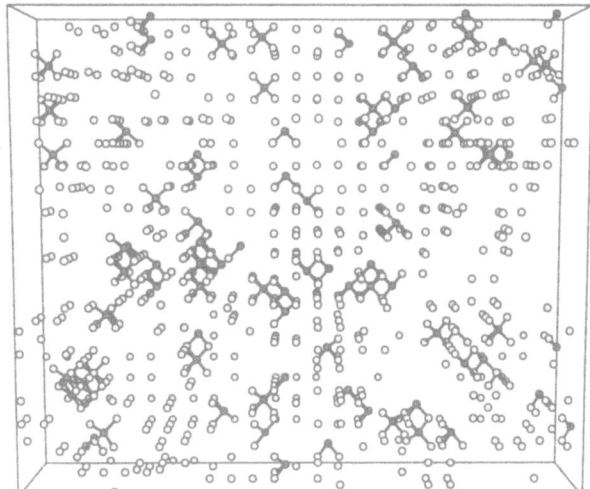

Fig. 4.24. Computer model of the equilibrium defect structure in $Fe_{0.92}O$ at $T = 1150°C$ based on measured short-range order parameters. Only few layers of the model are displayed. "Bonds" are drawn between interstitials and nearest-neighbor vacancies, while regular cations and anions are not shown. Typically, there is a large fraction of "unbound" vacancies, which is important for the cation conductivity, while all interstitials are bound to vacancies. ("Bonds" to defects beyond the surface of the box are not displayed)

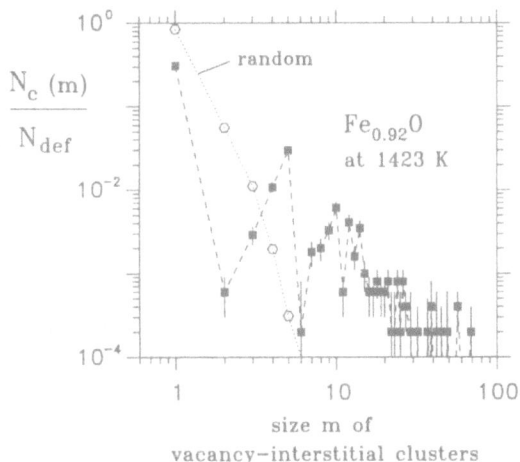

Fig. 4.25. Size distribution of vacancy–interstitial defect clusters (*filled squares*) in $Fe_{0.92}O$ at $T = 1150°C$ in comparison with the random case (*open circles*). The number of specific clusters $N_c(m)$ is normalized by the number of point defects N_{def}. More than 30% of the defects are isolated point defects (essentially Fe–vacancies) and 15% of the defects are bound in 4:1 vacancy–interstitial clusters. (Lines are only guides for the eye. From [4.46])

are interconnected by vacancy–interstitial "bonds" (this choice was justified by the large value of the related short-range order parameter) were identified for the cluster size distribution shown in Fig. 4.25. The remarkable features of the cluster distribution are that first, a large fraction of the defects, about 40%, are still unbound vacancies ($m=1$), which is important for the transport properties. Secondly, 4:1 clusters are particularly stable (note the log-scale), consistent with previous theories [4.51–4.53]. Thirdly, larger vacancy–interstitial aggregates, incorporating the 4:1 motif, are also present. Their existence appears to be, however, mainly a consequence of the high defect content. There is little or no tendency to form large clusters with well-defined compact shapes. Evidence of this is found from the observation that there is only very low intensity at small scattering vectors Q.

In comparison with the results for a random solid solution of equal composition, the characteristic features of the cluster distribution reveal that substantial and non-trivial many-body correlations were discovered from the measured full extent of the pair correlations.

This analysis does *not*, however, confirm any particular stability of other clusters which have been proposed in the literature. The likelihood of the 2:1 cluster as proposed by Roth [4.61], for instance, is less than 0.3%. Furthermore, no significance was found for the so-called Koch–Cohen 13:4 cluster, and also the 5:2 cluster proposed by Gartstein – which should consist of two incomplete corner-sharing vacancy tetrahedra aligned along ⟨110⟩ directions – could not be identified as typical of the defect structure. These defect clusters have been proposed from results obtained in earlier x-ray diffuse scattering experiments.

Instead of a clear dominance of a single cluster type only – note that 15% of all point defects are found in isolated 4:1 clusters – the distribution resembles a variety of local configurations, as is conceivable for the thermal equilibrium of a disordered high-temperature structure. In particular, the fraction of unbound vacancies offers an explanation of why the cation mobility is so high [4.55], $D \approx 10^{-7} \text{cm}^2/\text{s}$ at 1000°C, despite the strong correlation and binding between interstitials and vacancies. There is a peculiar behaviour of the diffusion coefficient, since it does not increase with increasing non-stoichiometry – at least for temperatures up to 1000°C [4.55]. Because of this, Monty [4.62] anticipated that "the diffusion coefficient could be proportional to the concentration of free vacancies in equilibrium with these aggregates of larger cluster following a complex law".

The case of dilute cation defects cannot be realized in FeO. Therefore, Hoser et al. [4.63] studied the equi-structural system iron-doped NiO. Indeed, the same type of diffuse scattering patterns were observed and their origin due to the Fe impurities was apparent. In this sense the iron-doped NiO may be regarded as a dilute wüstite. Kanzaki model calculations for the same type of defect arrangement, i.e. four cation vacancies surrounding an Fe–interstitial, yield a favorable description of the data. However, a mutual screening effect of these clusters leads to a slightly reduced Huang scattering even for defect concentrations as low as 1 %. This observation can be described by more general Kanzaki models as have been used for the FeO case, including regular ions as defects as well. Of course, in the asymptotic dilute limit these models make no difference.

Earlier investigations of the transition metal oxide $Mn_{0.936}O$ by Schuster et al. [2.31] have reported that cation vacancies with qualitatively the same displacement fields arising from the net Coulomb interaction should be the dominant defect type. (This paper did not take into account a proper Debye–Waller correction. However, the error seems to affect mostly the values of the displacements, which have been overestimated by approximately a factor of two, rather than the qualitative result of the dominance of uncorrelated vacancy defects.) More recent, unpublished results for larger non-stoichiometries [4.64] indicate, in agreement with powder diffraction data by Radler et al. [4.65], that indeed the presence of interstitials surrounded by cation vacancies cannot be denied, although preliminary estimates show that the fraction of vacancies should still be significantly larger than in the wüstite case. Such defect clusters of vacancies and interstitials have been proposed not only for FeO but also for the transition metal oxides NiO, CoO and MnO, by Catlow and Stoneham [4.51].

It is worth mentioning that Tetot et al. [4.66] have investigated the thermodynamic properties of interacting simple vacancy defects in models of these sodium-chloride structured transition metal oxides by Monte Carlo methods. They found that there is some need to go beyond the approximate models usually applied, which are based on the ideal mass action law formalism or on Debye–Hückel theory.

4.3.2 Short-Range Order and Interactions in the Metal Hydride VD_x

Metal hydrides have attracted a long-standing scientific and technical interest because of the unusually high hydrogen mobility and the high solubility of hydrogen. Vanadium hydride may be viewed as a model system for a compact solid hydrogen storage medium and technical applications of energy storage, although, in practice, there may be more suitable metal and alloy hydrides. At equilibrium the concentration of hydrogen in vanadium is determined by the temperature and the partial pressure of the surrounding hydrogen gas. With an increase in the H_2 pressure the hydrogen concentration in the metal hydride saturates, all available sites have been filled and almost as much hydrogen has been absorbed as there are metal atoms. A structural analysis by Bragg diffraction spots reveals that at sufficiently high temperatures the hydrogen has dissolved more or less randomly on the tetrahedral sublattice. Since this sublattice offers about six times more sites than those used, there are apparently strong correlations and repulsive interactions between the hydrogen atoms, which limit the storage capability of the material. As in previous examples, such correlations can be studied by diffuse scattering experiments in a very quantitative manner and compared with theoretical models. In addition to any direct or screened Coulomb interaction, it is of particular interest here to discuss also the role of long-range elastic interactions between the hydrogen atoms mediated by the metal host lattice. For an insight into the particular results for the vanadium hydride system, the problem of how to treat long-range interactions in Ising-type models will be briefly addressed in this context. Of course, this is not only of relevance for interstitial systems. There are only a few, recent Monte Carlo simulations of such model systems, e.g. for SiGe, and these will also be discussed with respect to general predictions for the statistical physics of systems with long-range interactions.

The dissolved hydrogen – it occupies preferentially the tetrahedral sites in bcc metals and the octahedral sites in fcc metals – has been interpreted in terms of a "lattice gas" [4.67, 4.68], assuming a relatively weak influence on the metal host lattice. The tetrahedral hydrogen sublattice is shown in Fig. 4.26.

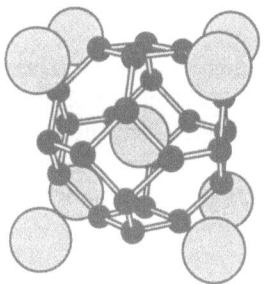

Fig. 4.26. The body-centered cubic structure of vanadium and the tetrahedral sublattice which is partially occupied by hydrogen

As shown for α-NbH$_x$ (Alefeld [4.68,4.69]), the H–H interaction is partially mediated by the long-range elastic strain fields, described by the interaction of elastic dipoles in a cubic medium. At short distances the strong repulsive Coulomb interaction of the screened protons is expected to lead to a mutual blocking of the first two neighboring shells. Blocking was also deduced from the measured solubilities [4.70], indirectly from the existing ordered phases [4.71, 4.72], and further from the wave-vector dependence of the chemical diffusion constant as seen in quasi-elastic neutron scattering [4.73]. Evidence of such a short-range order in VD$_{0.75}$ was also found by neutron diffraction of a polycrystalline sample [4.74].

Horner and Wagner [4.75] successfully described the gas–liquid-like phase transition in NbH using a hard core potential to mimic the real screened Coulomb interaction which was unknown in its details, and in addition, a long-range elastic term including the macroscopic contributions which result from the relaxation of a finite crystal. (It is this macroscopic relaxation of the finite crystal which causes the incoherent miscibility gap, typically occuring in decomposing, phase-separating alloys but also in heterogeneous ordering reactions. In contrast to homogeneous ordering, the transition temperatures related to coherent structural changes occur at comparatively lower temperatures and cannot be studied under equilibrium conditions.)

The scattering investigations of $VD_{0.781}$ (Pionke et al. [2.15, 3.80, 4.76]) were aimed at achieving with particularly high precision a determination of the real microscopic interactions in such a metal–hydrogen system.

From an experimental point of view it is a favorable model system, since the vanadium host was almost invisible to the neutrons (its coherent scattering length is negligibly small, $b_V = -0.0382 \times 10^{-12}$ cm). Therefore, the three-dimensional scattering experiments were performed in diffraction mode using the flat-cone diffractometer E2 at the Hahn–Meitner Institut, Berlin. The small inelastic contributions of the deuterium were taken into account by calculating the cross-sections of the optical and acoustic modes. Deuterium was chosen to avoid the high (spin-)incoherent background of natural hydrogen.

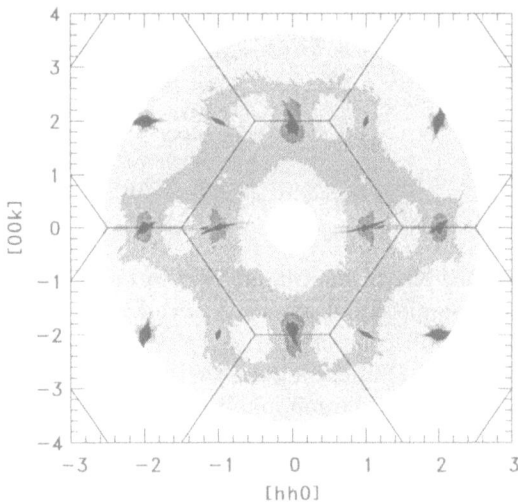

Fig. 4.27. Observed scattering of $VD_{0.781}$ in the (110) plane of the reciprocal lattice at $T = 300\,K$. The diffuse scattering in between the Bragg peaks is mainly due to short-range order among the deuterium interstitials. The translational symmetry properties for the tetrahedral sublattice are indicated in the figure (from [3.80])

The observed scattering in the basal (110) plane at room temperature is shown in Fig. 4.27. The symmetry properties of the tetrahedral sublattice are shown in the figure. Diffuse maxima are found near the 200 and 110 reciprocal lattice positions and equivalent positions which coincide with the Bragg peaks of the host lattice. The coherent diffuse scattering due to the deuterium short-range order is very low near the zone centers, which reflects the low compressibility of the deuterium "lattice liquid" at this high deuterium concentration.

The data were analysed using a Fourier analysis with respect to short-range order and displacement parameters, based on least-squares methods.

Figure 4.28 shows a strong blocking tendency on the tetrahedral interstitial sublattice for the first three neighboring shells of deuterium. There are only small modifications if the additional contributions from the octahedral sublattice are also taken into account. From Bragg diffraction data [4.77], and also consistent with the absolute calibration of the diffuse scattering, about 10% of the deuterium atoms in this study were observed to be distributed on the octahedral sites.

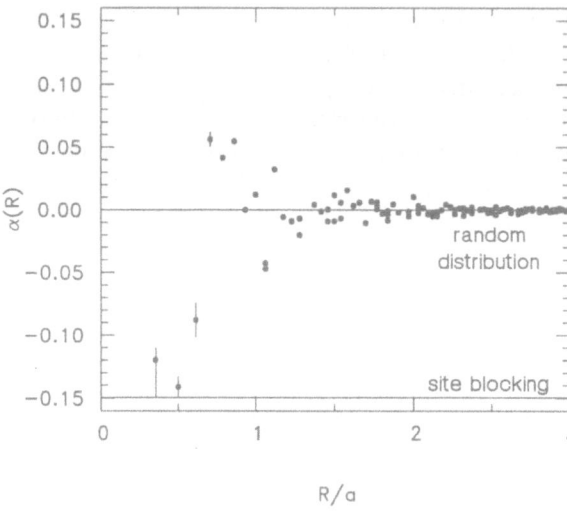

Fig. 4.28. The short-range order parameters for deuterium on the tetrahedral sublattice of $VD_{0.781}$ in thermal equilibrium at room temperature (a=lattice constant). A strongly reduced occupation probability is found for the first three neighbor shells of deuterium (from [3.80])

The mean-field approach using the Krivoglaz–Clapp–Moss formula to the D-D interaction energies would be completely insufficient, since for the near-neighbor pairs V_{D-D} is not at all small compared with $k_B T$. In addition, there is a breakdown of the usual MF prediction that the wave vector of the short-range order peak corresponds to the minimum of the interaction potential $V(q)$.

According to our results obtained by the inverse Monte Carlo method from the diffuse scattering data (Fig. 4.29), typically all the interactions determined from the experiment have a positive sign. This could be attributed to the repulsive part of the interaction due to the (screened) Coulomb interaction between the protons and shows that the Coulomb term is dominant within the whole displayed range of interactions.

On the other hand, the elastic interactions displayed for comparison in Fig. 4.29 are typically below the inverse MC result, but they are of comparable magnitude and the variations with R of both data sets reveal much similarity. These continuum-mechanical calculations of the elastic interaction were performed by Siems [4.78] and took account of the dipole–dipole interaction. One has to distinguish the possible relative orientations of the dipoles, for instance in $\langle 110 \rangle$ directions the dipoles are oriented alternately perpendicular

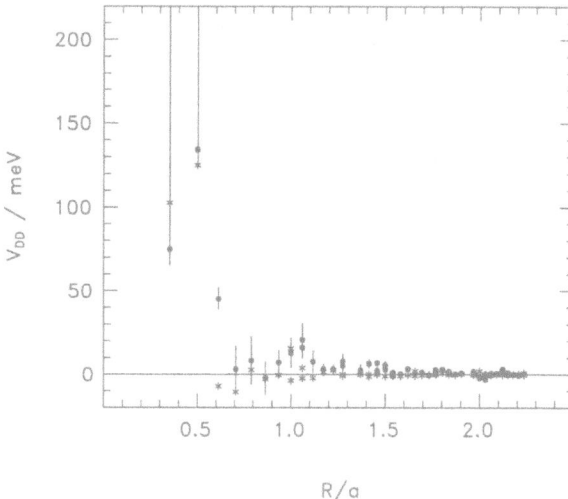

Fig. 4.29. D–D interaction parameters V_{D-D} determined from short-range order parameters using the inverse MC method (*full circles*) compared with continuum-mechanical calculations of the elastic interaction (*asterisks*) (from [3.80])

and parallel, and further the projections of the dipoles on $\boldsymbol{R}$, which give for instance two different results for pairs at distance $\boldsymbol{R} = \frac{1}{4}(4,0,0)\,a$. However, the surprising agreement of the continuum-mechanical calculations with the experimental result at very close distances seems to be more accidental than really meaningful. The calculations may give a good approximation for the observed values at larger distances. Calculations of the stress-induced interaction between interstitial defects in bcc solutions were performed by Blanter and Khachaturyan [4.79], who took into account the discrete lattice properties (Born–von Karman model). A comparison of their results with those for the near distances does not show any similarity; in fact they differ in sign. If their results are correct, the contradiction may indicate that at short distances the elastic contributions are indeed not very important compared with the repulsive screened Coulomb interaction between the protons.

These experimentally determined D–D interaction results may serve to test various theoretical calculations. For instance, such theoretical predictions as [4.80] which consider only the electronic part of the interaction are severely insufficient to describe the experimental results. Furthermore, it appears to be a challenge to compute the coherent phase diagram of the V–D alloy system by Monte Carlo methods on the basis of these realistic interaction parameters. This may be just feasible in view of the current computing power, despite the large number of interactions.

It is worthwhile to spend a few words on long-range, attractive elastic interactions which are also present in $\mathrm{VD_x}$, because of the anisotropy of both the defect dipole tensor [4.72] and of the host lattice (for a general introduction see, for instance, Leibfried and Breuer [4.81]). There are attractive and repulsive directions, which will affect the morphology of the decomposing structure (see Chen et al. [4.13]), while upon orientational averaging, the dipole–dipole interaction is equal to zero.

Such cases should exhibit mean-field exponents in the vicinity of the (coherent) critical temperature. Of course, it is not true that interaction models with elastic interactions lead to mean-field exponents in any case or that mean-field treatments are quantitatively correct. For one thing, this may hold only for critical decompositional fluctuations (for $q_c = 0$), and not for continuous ordering transitions with $q_c \neq 0$.

Interactions which decay as $1/r^{d+\sigma}$, where d is the dimensionality of the system and $\sigma \geq 0$, are called *short-range* for $\sigma \geq 2$ and *long-range* for $\sigma < 2$. As shown by Sak [4.82], for $2 - \eta < \sigma < 2$, the exponents (for the static critical properties) of these long-range interactions assume their short-range values. According to Fisher et al. [4.83], for $0 < \sigma < 1.5$, the exponents α, β, γ are classical, while $\nu = 1/\sigma$ and $\eta = 2 - \sigma$ depend on the power of the r dependence of the interaction potential. (The border with the classical regime for the dynamic critical properties does not coincide with that for the static properties, as recently analyzed by Folk and Moser [4.84].)

Elastic interactions between point defects arise from the anisotropy of either the medium or the defects themselves. The dipole–dipole type of elastic interaction (leading term $\propto r^{-3}$) cancels for crystals with isotropic properties in the long wave limit (i.e. equal shear constants: for cubic crystals $c_{44} = \frac{1}{2}(c_{11} - c_{12})$).

For instance, Al–Zn solutions at about 40 at% Zn are very nearly elastically isotropic [1.1]. Therefore, the critical fluctuations may be expected to show the Ising exponents rather than the mean-field exponents observed by Schwahn et al. [4.85]. Experimental investigations are not so easy or unambiguous, since the incoherent decomposition interferes and makes the direct experimental observation difficult, as well as a precise determination of the spinodal temperature.

We will briefly discuss the role of the elastic energies involved in the decomposition in alloys, and the related approaches by Monte Carlo methods which have found increasing interest in recent years. In alloys of Si–Ge semiconductors there is a typical "size effect". At ambient temperatures there is a solid solution possessing the diamond structure (which lacks inversion symmetry). The alloy has been predicted to decompose at lower temperatures by de Gironcoli et al. [4.86], where equilibrium can hardly be attained in experimental studies. The chemical decomposition has been the subject of various Monte Carlo simulational studies, exemplifying the possible methods of including elastic interactions. One possible approach is to determine the effective Ising type and long-range pair interaction

$$J_{mn}^{\text{eff}} = J_{mn}^{\text{chem}} - \sum_{kl} f_{mk} \Phi_{kl}^{-1} f_{ln} \tag{4.3}$$

by which the configurational statistics can be treated on a rigid lattice within the assumptions of harmonic lattice properties, Kanzaki forces related to mere size effects, an average lattice Green's function and the linear super-position of the displacement fields. The disadvantage of such an approach is that the effective interactions are of long range (though the forces f_{ik} are typically of short range only), so that in applications of this method the effective interactions were truncated to finite range for practical reasons. This cannot reproduce the correct asymptotic critical behavior but is certainly a reasonable and sufficient approach for phase diagram calculations. Examples also exist for the Si–Ge alloy (de Gironcoli et al. [4.86] and Wolverton and Zunger [4.87]).

By including the positional degrees of freedom in the MC simulations one can avoid, if necessary, the approximations mentioned above. In this context, we again draw attention to the model calculations by Chakraborty [3.16], who included the effects of the displacements on the chemical interactions. Furthermore, such a direct approach appears to be attractive with respect to the computational effort, since the origins of the size effects are usually only short-range forces or potentials. The recent Monte Carlo simulations by Dünweg et al. [4.88] using the Keating potential determining bond length and angles may serve as an example. This indeed yielded mean-field exponents ($\gamma = 1$) close to the critical temperature. In a following, similar Monte Carlo study of Si – –Ge by Laradji et al. [4.89], the better-suited Stillinger–Weber potential was used and again mean-field critical exponents were found. These results might have been anticipated, since the effective interactions J_{mn}^{eff} belong to the dipole–dipole type $\propto r^{-d}$, where no cancelation of the asymptotic long-range part should occur in the case of an anisotropic dipole force tensor and/or anisotropic elastic properties of the underlying lattice.

The recent work of Vandeworp and Newman [4.90], simulating a binary alloy on a square lattice with a Keating-like potential and continuous positional degrees of freedom, is also recommended as a readable introduction to the topic. They found that for an ordering binary alloy, transition temperatures were raised owing to the elastic part of the interaction, while it was argued that for decomposing alloys the coherent transition temperatures are lowered. The incoherent phase boundaries will be raised in temperature in comparison with the pure chemical interaction model, as seen in the simulations by Dünweg et al. and Laradji et al. [4.88, 4.89].

One may note further that the findings of Vandeworp and Newman, namely that their model belongs to the two-dimensional Ising universality class, cannot be taken as an argument that effective long-range interactions do not exist. The reason is simply that the ordering wave vector q_c is not equal to zero but refers to a zone boundary mode of highest symmetry. Therefore the long-range part of the effective interaction cancels, and the ordinary (Ising) behavior will be recovered, as usually seen in experimental studies of critical phenomena in ordering alloys.

4.4 Surface-Induced Ordering and Disordering near First-Order Bulk Transitions

The previous sections have given a survey of what kind of information about the short-range order and effective interactions can be inferred from diffuse scattering experiments. These results were all related to bulk properties. Now we turn to the effects of surfaces on microstructure, in particular close to the order–disorder transitions. In this section Monte Carlo results [4.91, 4.92] for a most simple but generic model will be discussed rather than experimental studies.

A variety of complex, new phenomena occur at interfaces which cannot be understood from the properties of the bulk or surface only, e.g. the physics of related phenomena such as heterogeneous nucleation, melting, adsorption and wetting (for reviews of wetting phenomena see [4.94]). The scenario of wetting requires three distinct phases, for instance vacuum, the bulk phase and a third phase intervening in between at equilibrium. As observed in the situation of melting [4.95], the transition at the surface may be continuous even in the case of first order bulk transitions, a particular situation which has been theoretically investigated by Lipowsky [4.96]. In the analogous case of surface-induced disorder – which is one topic to be discussed in detail in this section – a film of disordered layers "wets" the bulk phase as the bulk transition temperature is approached. For first-order bulk transitions, according to renormalization group methods, the critical behavior is in some respects correctly described by the mean-field theory; some of the exponents are, however, even non-universal and will depend on the interfacial roughness of the specific material considered [4.96, 4.97]. Because of the valid scaling relations, there is only a single independent critical exponent (see [4.94]), which can be related to the interfacial stiffness $\tilde{\Sigma}$. For instance, the order parameter of the surface layer is $\beta_1 = 1/2 + \omega/2$, for $0 \le \omega \le 1/2$ which is the case of interest here, where $\omega = k_B T_{cb}/(4\pi \tilde{\Sigma} l^2)$, T_{cb} is the bulk transition temperature and l is the intrinsic thickness of the interface. Below the roughening temperature, which depends on the specific interface, $\omega = 0$ and the non-universal exponents such as β_1 take their mean-field values. In this case, the interface should jump from layer to layer as T approaches T_{cb}, and layering transitions rather than a continuous wetting process could be observed [4.96, 4.98] (which may still be approximated by effective exponents in a log–log plot). Independent of any interfacial roughness, the interface should proceed into the bulk according to a logarithmic law $\propto \ln(1/t)$ [4.96]. Besides the reviews by Lipowsky [4.96] and Dietrich [4.94] a recent review by Binder [4.99] gives an excellent introduction to the subject.

Experimental approaches by scattering investigations have been made using probes which are more sensitive to surface properties than neutrons. Ion beam scattering and low-energy electron diffraction (LEED) [4.100] are particularly sensitive to the top surface layer, and x-ray scattering in total reflection is well suited to determine the near-surface profile within the interesting length scale in the range 1 Å to about 100 Å and more. A nice review of such experimental investigations has been given by Dosch [2.44], among these being studies of surface-induced disorder in Cu_3Au. Most studies are concerned with the new critical phenomena near the surface, measuring the profiles of the relevant order parameter but also the critical diffuse scattering. The off-specular diffuse scattering is caused by the spatial ordering correlations within the various layers parallel to the surface or interface; e.g., as measured for Cu_3Au, the intensity around (110) measured in transmission is bulk sensitive and in reflection it is surface sensitive [4.101]. For second-order bulk phase transitions, the near-surface critical scattering (from NH_4Br) yielded the universal critical exponents $\beta_1 = 0.8$, $\eta_\| = 1.3$ and $\nu = 0.5$, which differ from the two-dimensional as well as the bulk three-dimensional values [4.102]. For first-order bulk transitions the best-supported result from experiments is a logarithmic growth law for the thickness of the wetting layer as T_{cb} is approached, as has been theoretically predicted [4.96].

Wetting of anti-phase boundaries has been observed in various alloys by TEM investigations, e.g. [4.103]. While at the surface one may expect important relaxations, even surface reconstructions and surface effective fields (at least because of missing bonds at a free surface, see [4.99]), the interfacial properties of anti-phase boundaries (APBs) can be related to the effective pair interaction derived from diffuse scattering experiments on bulk systems. Agreement was found between the observations of the APB width and the dislocation dissociation versus temperature in (weak beam) transmission electron microscopy and cluster variation method model calculations based on such experimentally determined interactions. Kikuchi and Cahn 1979 [4.104] predicted such wetting phenomena at APB's and made detailed calculations of interfacial profiles. Suttle layering transitions were found in this model later by Finel [3.5]. At the present there is no investigation which has attempted to analyze the possible modifications of the interactions near the surface by the analogous approach of using the inverse MC analysis of the diffuse scattering outlined in the previous sections.

Monte Carlo simulations of simple model systems are helpful for studying the generic behavior of phase transitions near the surface and to check the detailed theoretical predictions, in particular the scaling predictions of the actual wetting theory. To get access to the critical region, however, one has to perform long simulations on extremely large systems, as was done in a recent investigation for an fcc alloy model by Schweika, Landau and Binder [4.91, 4.92]. For a proper analysis of the specific near-surface behavior, we needed to determine very accurately the first-order bulk transition temperature $T_{cb} = 1.738005(25)|J|/k_B$ for the nearest-neighbor interaction model. The Monte Carlo study [4.91, 4.92] considered this simplest model for a CuAu type of alloy – the nearest-neighbor antiferromagnetic Ising model on an fcc lattice. Since the nearest-neighbor geometry implies nearest-neighbor triangles, an antiferromagnetic type of order is not possible without *frustra-*

tion. Several peculiar aspects of the bulk ordering can also be related to this fact. For instance, no continuous, but only first-order, (bulk) transitions are usually observed. The transition temperatures are low ($kT_c/z|J_1| = 0.1448$ for $J_1 < 0$, compared with $kT_c/q|J_1| = 0.816$ for $J_1 > 0$); long-period structures may occur due to small APB energies. There is an unusual temperature dependence of the short-range order, and surface-induced ordering can occur at (100) free surfaces due to the geometrical constraints rather than only surface-induced disordering because of missing bonds. All these properties arise from the competition between the three ordering variants which are alternate layerings of pure "Cu" and pure "Au" planes in one of the three $\langle 100 \rangle$ directions. With the four simple cubic sublattice "magnetizations" for the fcc lattice (see Fig. 4.30), we have

$$\psi = \begin{pmatrix} \psi_x \\ \psi_y \\ \psi_z \end{pmatrix} = \frac{1}{4} \begin{pmatrix} m_1 - m_2 - m_3 + m_4 \\ m_1 - m_2 + m_3 - m_4 \\ m_1 + m_2 - m_3 - m_4 \end{pmatrix}.$$

However, a scalar order parameter ψ based on the occupations of only two sublattices appears to be a proper choice if one variant of ψ dominates. In such cases, and these will be the typical ones here, ψ_n denotes the order in a single layer n, while $\psi_{x,n}$ etc. always refers to a bilayer.

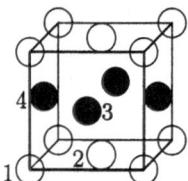

Fig. 4.30. CuAu structure ($L1_0$)

4.4.1 Surface-Induced Disorder

A free (111) surface lowers the ordering energy in the surface layers, and hence below T_{cb} one should expect a lower degree of order near the surface in comparison to the bulk (see Fig. 4.31). The actual ordering variant is determined by the bulk order, and a scalar order parameter appears to be sufficient. The interfacial profile can be fitted well to

$$\psi_n = \psi_{\text{bulk}} \{ 1 + \exp[-2\xi_\perp^{-1}(n - \hat{n})] \}^{-1} \tag{4.4}$$

which is analogous to a tanh function (which Cahn and Hilliard [4.105] had derived in 1958 to describe an interface between two ordered domains at a second-order bulk transition). The thickness of the wetting layer $\bar{l}$ and the excess surface order $\psi_s = \sum(\psi_n - \psi_{\text{bulk}})$ are also indicated in Fig. 4.31. There are slight modifications of the profile to be seen near the surface. According to Boulter and Parry [4.106], precisely these modifications could introduce an important length scale which ought to be incorporated in the theoretical description of wetting.

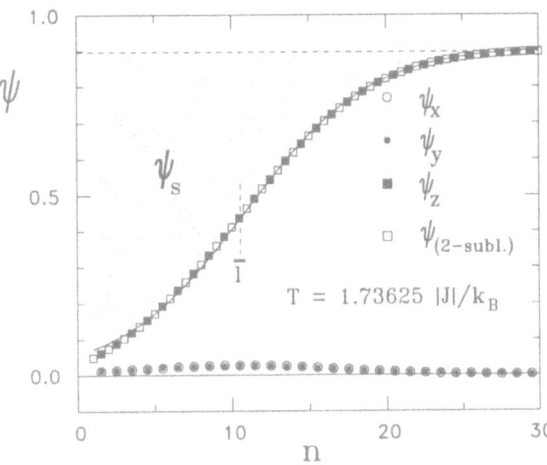

Fig. 4.31. Order parameter profile for layers n at the (111) surface slightly below T_{cb}. ψ_{s} denotes the (integral) excess order near the surface. The position of the interface between the disordered near-surface region and the bulk is also a measure of the thickness $\bar{l}$ of the wetting layer

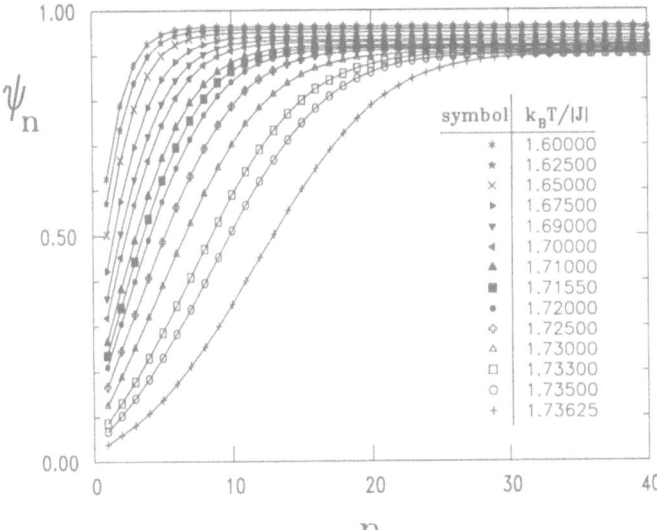

Fig. 4.32. Profiles of the layer order parameter ψ_n for the (111) surface plotted versus layer number n and for temperatures close to the first-order bulk transition

Figure 4.32 shows how the interface proceeds into the bulk as the temperature is raised close to the first-order bulk transition at T_{cb}. Figure 4.33 illustrates the logarithmic growth law for the disordered layer thickness $\bar{l}$ as the bulk transition temperature is approached.

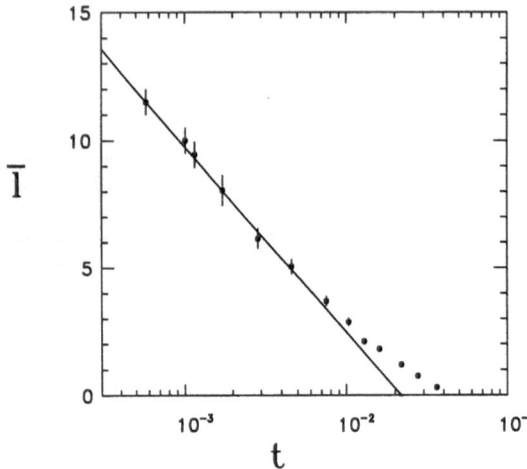

Fig. 4.33. Logarithmic divergence of the disordered layer thickness $\bar{l}$ versus the reduced temperature $t = 1 - T/T_{cb}$.

The excess quantities of order $\psi_s = \sum(\psi_n - \psi_{bulk})$ and energy E_s show the predicted logarithmic divergence, i.e. $\beta_s = 0$. In addition, a predicted root-logarithmic law for the width of the interface [4.97] was verified by the Monte Carlo results [4.91] (see Fig. 4.34). Details of the proportionalities, however, disagreed with the theoretical predictions [4.97].

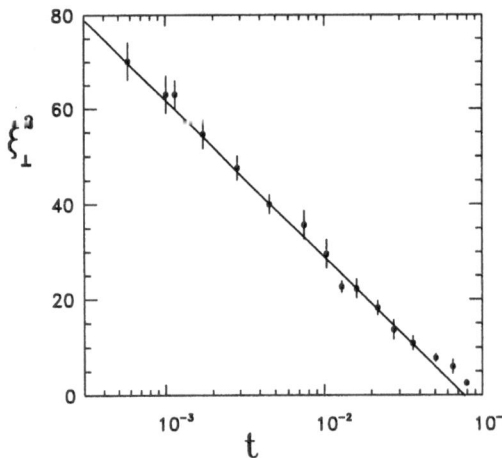

Fig. 4.34. The interfacial thickness exhibits a root-logarithmic divergence as $t \to T_{cb}$

The first layer shows a continuous transition at T_{cb} with a non-universal exponent $\beta_1 = 0.64$. Figure 4.35 also demonstrates that the subsequent layers order with the same exponent but with increasing critical amplitudes.

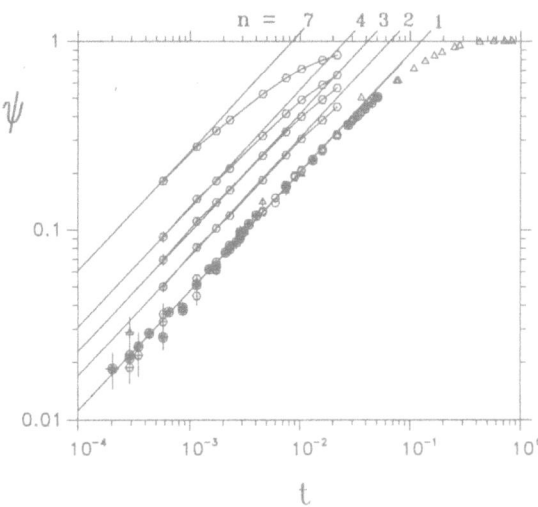

Fig. 4.35. Log–log plot of the layer order parameters versus $t = 1 - T/T_{cb}$ at a (111) surface. The same exponent $\beta_1 = \beta_n = 0.64$ but different amplitudes are found for the ordering in the surface and in subsequent layers

For a further, more crucial test, derivatives of the order parameter and of the surface excess energy were analyzed. Three susceptibilities were considered: $\chi_{1,1} = \partial\psi_1/\partial H_1$, i.e. the derivative of the order in the first layer with respect to a conjugate (staggered) surface field H_1 (and more precisely the singular part of it, which is the difference from the susceptibility in the presence of a disordered bulk phase); $\chi_1 = \partial\psi_1/\partial H$ and $\chi_s = \partial\psi_s/\partial H$, where H is the field conjugate to the order and acts on the whole crystal. Since there is only one independent critical exponent, which is further related to the interfacial stiffness, the asymptotic behavior for $t \to T_{cb}$ of these quantities is expected to be consistent with the following scaling relations for the corresponding critical exponents [4.91]:

$$\beta_1 + \gamma_1 = 1, \tag{4.5}$$
$$\beta_s + \gamma_s = 1, \tag{4.6}$$
$$\gamma_{1,1} + \gamma_s = 2\gamma_1, \tag{4.7}$$
$$\gamma_{1,1} + 2\beta_1 = 1. \tag{4.8}$$

The data for these susceptibilities and the surface excess specific heat C_s, shown in Fig. 4.36, were obtained from the appropriate fluctuation relations. The straight lines, which are shown for comparison, result from theoretical predictions and scaling relations using $\beta_1 = 0.64$. The asymptotic region seems to be reached rather slowly, in particular for χ_s. (This may be due to logarithmic corrections according to the log-divergence of ψ_s, since $\beta_s+\gamma_s = 1$ is a valid scaling relation, and β_s tends to zero only asymptotically.)

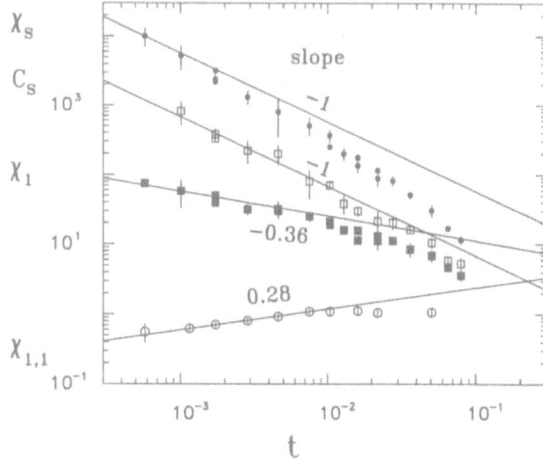

Fig. 4.36. Log–log plot of the excess specific heat C_s and the various susceptibilities versus $t = 1 - T/T_{cb}$. Straight lines indicate the exponents, $\gamma_s = 1$, $\alpha_s = 1$, $\gamma_1 = 0.36$, $\gamma_{1,1} = -0.28$, which are theoretically expected from the scaling relations and β_1

Surface-induced disorder is not an exceptional case for the free (111) surface of this model. Other facets, where the surface reduces the ordering energy, for instance those with normal directions $\langle 110 \rangle$, are expected to exhibit the same qualitative phenomena.

4.4.2 Surface-Induced Order due to Reduced Frustrations

While the bulk ordering of the CuAu-type model is unavoidably frustrated because of geometrical constraints, this is not the case for a 2×2 superstructure within the (100) layers which form this structure. Frustration (in the case of the nearest-neighbor interaction model) occurs only between such ordered layers. Hence, the removal of a neighboring layer by a free surface should stabilize the order in the surface layer. In fact, the transition of the purely two-dimensional system, i.e. the square lattice, takes place at a higher temperature ($T_c^{sq} = 2.27|J|/k_B$) than the transition of the nearest-neighbor fcc model. Therefore, surface-induced ordering may occur in CuAu-type alloys without any enhancement of the interactions near the surface (which might nevertheless be realistic for some alloys), as shown in the first Monte Carlo simulations (of an interaction model including first and second nearest-neighbor interactions) by Schweika, Binder and Landau [4.107]. In a recent study by these authors [4.91–4.93] of the nearest-neighbor interaction model, surface-induced ordering was investigated in detail.

While surface-induced ordering and surface-induced disordering are commonly described in the literature as merely equivalent phenomena, one has to note this difference, that for surface-induced ordering the surface necessarily orders at a different and higher temperature than the bulk. One important consequence is that the predictions for the surface properties, and also the scaling relations mentioned above (see Figs. 4.35 and 4.36), can no longer be valid. Instead, at the surface transition one expects a behavior which falls in the universality class of the corresponding purely two-dimensional system. Therefore, the surface order parameter vanishes continuously as the surface transition temperature is approached as $\psi_1 \propto (T - T_{cs})^{-\beta_1}$ where β_1, however, is the well-known two-dimensional Ising exponent $\beta^{2D} = 1/8$. On the other hand, for $T \rightarrow T_{cb}$, the logarithmic law is still expected to provide a valid description for the temperature dependence of the excess quantities, e.g. those of the order, the energy and the interface position. Figure 4.37 shows an order parameter profile slightly above the bulk transition at T_{cb}. Here, the situation is reverse, when compared with the (111) surface, and order is induced near the surface in the presence of a disordered bulk. Actually, two ordering variants of the CuAu structure are degenerate (ψ_z and ψ_y) and may compete with each other, which complicates the situation as discussed below; the third variant (ψ_x), which corresponds to an alternate layering of pure "Cu" and "Au" planes, is unstable. As already seen in the case of the (100) surface, the tanh function describes the interfacial profile quite well apart from small deviations near the surface.

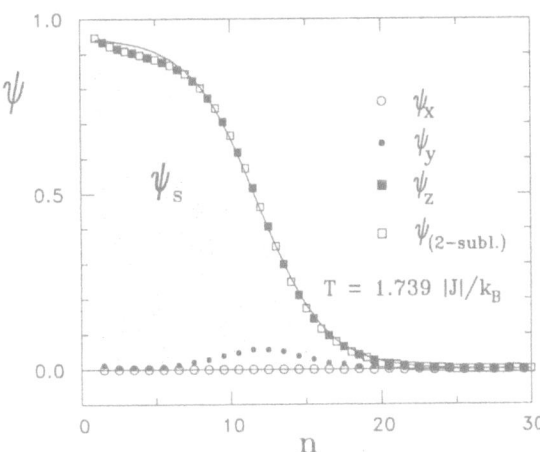

Fig. 4.37. Order parameter profile for layers n at the (100) surface slightly above T_{cb}

Next, Fig. 4.38 shows how the interface moves from the surface into the bulk as T approaches T_{cb} from above. The profile for a temperature closest to the surface transition at T_{cs} indicates an interesting detail. It can be seen that the order in the first plane has already induced order in the third layer, while the second layer is still disordered. Indeed, there are two transitions in the near-surface region above T_{cb}. At $T_{cs}^{(1)} = 1.896(1)|J|/k_B$, ordering takes place in the first layer and with exponentially decreasing amplitudes in subsequent odd layers. At $T_{cs}^{(2)} = 1.791(3)|J|/k_B$ the system decides to choose either the order parameter ψ_x or ψ_y, and the order–disorder transition occurs in the second and subsequent even layers. With finite-size scaling methods, we determined the transition temperatures $T_{cs}^{(1)}$ and $T_{cs}^{(2)}$ and β and ν were

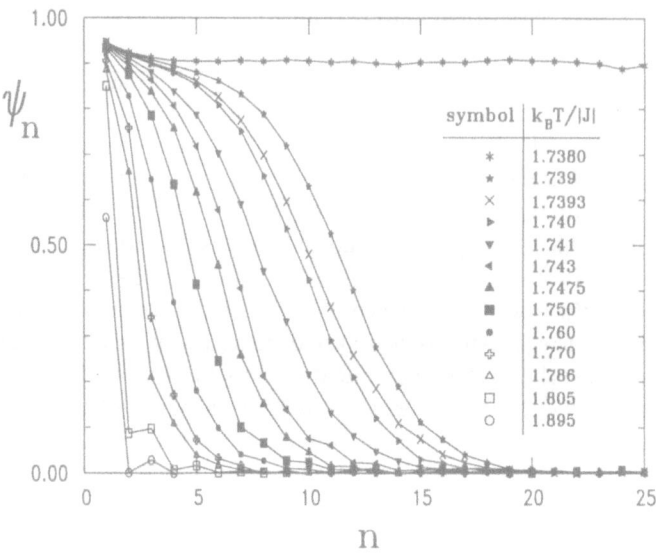

Fig. 4.38. Profiles of the layer order parameter ψ_n $(= \psi_{2-\mathrm{subl}})$ for the (100) surface plotted versus layer number n and for temperatures close to the first-order bulk transition

found to be in good agreement with the two-dimensional values as already mentioned [4.91].

Considering quantitatively how the interface penetrates into the bulk, as $T \to T_{\mathrm{cb}}$, a logarithmic divergence was found for the excess quantities of order and energy, the interface position and also the squared interfacial width, as theoretically expected for this wetting situation (see Fig. 4.39).

4.4.3 Surface-Induced Order due to Effective Surface Fields

In real alloys, it is likely that there are more surface effects than only missing bonds. For instance, there may be effective surface fields leading to the enrichment of one component in the top surface layer. The origin of such fields is due to another phase, where the vacuum appears only as a special choice of possible adsorbates (solid-on-solid models). As pointed out by Binder [4.99, 4.108] effective surface fields arise as soon as the interactions V_{AA} differ from V_{BB}, while breaking this symmetry has no effect on the bulk properties of a pairwise Ising Hamiltonian.

Furthermore, one could expect that the effective interactions are modified close to the surface. In particular, alloys of the real CuAu system are known to exhibit important atomic size effects [1.11, 4.109], and relaxations at the surface are expected to modify the on-site energy as well as the interaction energies. Both aspects have been taken qualitatively into account by Kroll

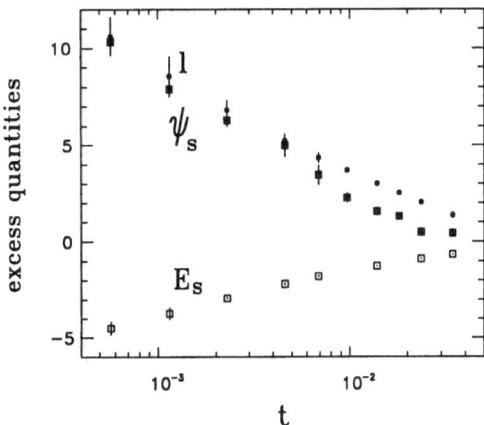

Fig. 4.39. Logarithmic divergence of the excess quantities l, ψ_{s}, and E_{s} versus $t = 1 - T_{\mathrm{cb}}/T$

and Gompper in theoretical investigations [4.110, 4.111] of surface-induced disordering at (100) surfaces of Cu_3Au alloys. In the present model and for a (100) surface the effects of surface fields are more drastic than for the Cu_3Au model. Here, the *surface field acts as a conjugate field* to an order parameter which is usually suppressed. A likely preference of, say, Au at the surface favors a layering AuCuAuCu... perpendicular to the (100) surface (characterized by ψ_x) and, if sufficiently strong, this can be expected to destabilize the other two cubic ordering variants, ψ_y and ψ_z, which prefer the (2×2) ordered surface.

This situation is illustrated in Fig. 4.40. Indeed, at a (100) surface of our CuAu-type model a sufficiently strong surface field favors a layering AuCuAuCu.. perpendicular to this surface. This is described by the variant $\psi_{x,n}$.

Unlike the previous cases of surface-induced ordering and surface-induced disordering, here, owing to the presence of a surface field which couples to the relevant order parameter ψ_x, there is no surface phase transition. Up to temperatures far above the bulk transition temperature T_{cb}, there exists an alternating concentration profile (which corresponds to ψ_x). Continuous wetting for $T \to T_{\mathrm{cb}}$, however, can only be expected for sufficiently strong surface fields which can increase the order parameter ψ_1 near to the value of ψ_{bulk} at T_{cb}. This seems to occur only for $h_1 \geq 3|J|$. Otherwise, only a finite thin film of order appears at the surface above T_{cb} (prewetting). For sufficiently weak surface fields – according to the Monte Carlo data for h_1 below approximately $1|J|$, while a ground state analysis would yield too high an estimate of $h_1^c = 2|J|$ – the previously discussed surface-induced ordering due to reduced frustrations survives.

More generally, in order to have such alternating concentration profiles, and possibly surface-induced ordering in the wetting sense, the effective surface field needs to couple to one variant of the bulk order parameter. This

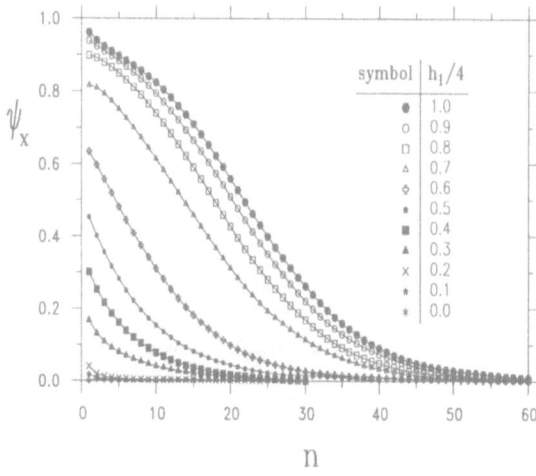

symbol	$h_1/4$
●	1.0
○	0.9
□	0.8
△	0.7
◇	0.6
■	0.5
■	0.4
▲	0.3
×	0.2
✳	0.1
✶	0.0

Fig. 4.40. Dependence of the order parameter profiles ψ_x on layer number n for a surface field h_1 slightly above T_{cb}, $T = 1.74\,|J|/k_B$

has been discussed recently by Schmid [4.112] in some more detail with respect to the FeAl and Cu_3Au alloys. Reichert et al. [4.113] measured by x-ray scattering experiments the concentration profiles in Cu_3Au above the order–disorder transition, and recently these were studied theoretically by Mecke and Dietrich [4.114] within a mean-field approach. The concentration profiles perpendicular to a free (100) surface were found to decay exponentially in an oscillatory manner towards the bulk value. In good quantitative agreement with experiment, this mean-field treatment showed that the decay length increased as $(T - T_{sp})^{1/2}$ upon approaching the first-order transition at T_c, where $T_{sp} < T_c$ denotes the spinodal temperature of the order–disorder transition. In contrast to the CuAu case, in Cu_3Au the effective surface field couples only to a subspace of the order parameter – favoring Au to be in the near-surface layer – and only a finite near-surface region is affected by this type of order at T_{cb} (incomplete wetting). A "staggered" field, which is of course not very realistic, would distinguish sufficiently the sublattice occupations for the Cu_3Au type of order, so that surface-induced ordering at T_{cb} (complete wetting) could occur as in the CuAu model discussed here. Our own preliminary Monte Carlo simulations, however, showed that the specific choice of the effective surface fields in the first two layers have a rather sensitive effect on whether surface-induced disordering occurs or not. Surprisingly, surface-induced ordering may even take place for alloys with the Cu_3Au type of order without any need for a "staggered" field. In particular, one may expect surface-induced ordering at the Au_3Cu composition.

Finally, it should be noted that the term *surface segregation* which is frequently used in this situation is rather misleading. The *surface enrichment by one component* is due to a field and not due to the atomic interactions. Therefore, as we are considering bulk ordering alloys, the states at the surface will typically be homogeneous, and the short-range order in the surface will be determined by the ordering tendency of the alloy.

A summary of the generic order–disorder behavior at the facets of CuAu-type alloys is given in Fig. 4.41. In addition to the temperature dependence of the order in the first layer for three distinct cases, the order parameters of the second and seventh layer are displayed to show how the discontinuous character of the bulk transition is approached within the deeper layers.

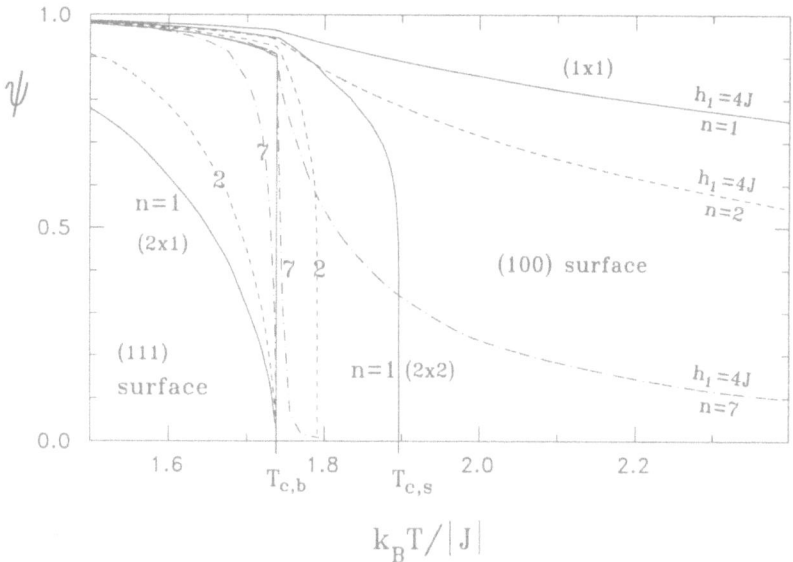

Fig. 4.41. Order parameter variation with temperature at the surfaces of a CuAu alloy model. Three distinct cases are shown: (i) surface-induced disorder at a (111) surface, $\psi = \psi^{(2\times 1)}$. There is a continuous transition of the surface layer, as well as of the subsequent layers at the discontinuous bulk transition T_{cb}. (ii) surface-induced order at a (100) surface, $\psi = \psi^{(2\times 2)}$, (without external fields). The continuous transition of the surface layer is at T_{cs}, i.e. above T_{cb}. The first and second layer order continuously $\propto t^{\beta}$, with $\beta = 1/8$, the value for the two-dimensional Ising universality class. (iii) surface-induced order at a (100) surface and a surface field $h_1 = 4J$ (which favors eithers "Cu" or "Au" in the surface layer), $\psi = \psi^{(1\times 1)}$.

One may also note the temperature dependence of $\psi_{x,n}$ (equal to $\psi^{(1\times 1)}$) for the surface field $h_1 = 4|J|$ and a (100) surface in Fig. 4.41. This case also refers to a situation where the surface is completely covered by one component, say Au, which induces an effective field of $4|J|$ on the next layer. Therefore, the result contains the limiting case of an infinite surface field – one only needs to shift the order profile by one layer – and exemplifies its relevance for adsorbates.

What is the relevance of these model calculations to the real CuAu alloy system? First, the effective pair interactions extend to further than only nearest or next-nearest neighbors, according to results from diffuse scattering experiments on Au_3Cu [4.116] and on Cu_3Au [3.35, 4.117]. At the CuAu stoichiometry, there appears an intermediate long-period ordered structure upon cooling (CuAuII structure), which confirms that the effective pair interactions are of longer range, presumably owing to Fermi surface effects. Nonetheless, for the same reason one has to conclude that the system is close to the degenerate situation, as found for the nearest-neighbor interaction model. Second, the surface field h_1 can be estimated from the partial vapor pressures or from the heats of formation. For the pure elements Cu and Au we have 80.7 kcal/mol and 88.0 kcal/mol respectively [4.115]. Therefore, the effective surface field due to the asymmetry of bulk interactions is about $h_1 \approx 2.5J$ (< 0), and its sign would say that Cu is preferred in the top layer. However, since this is at variance with the observed Au enrichment at the (100) surface of the real Cu_3Au alloy, one may speculate that the relaxational energies are important and push the larger Au atoms onto the surface. Therefore, with respect to CuAu, one would not expect that case (ii) of Fig. 4.41 with (2×2) order applies to real CuAu. Instead of this, it seems to be more likely that there will be ordering near the surface above the bulk transition of the ordered CuAu phase similar to case (iii), an AuCuAuCu... layering perpendicular to a (100) surface.

As already mentioned, Gompper and Kroll [4.111] have made an effort to simulate surface-induced disordering at the (100) surface of Cu_3Au. Recent Monte Carlo studies (Schweika and Landau [4.118]) of this alloy model reveal that their simple model is, however, unable to provide a consistent description of the observed surface-induced disorder *and* the near-surface segregation profile. A better agreement might be achieved from a more realistic interaction model [3.35] derived from the measured bulk short-range order [4.117], which shows that interactions to greater distances than the first and second neighbors are not negligible. One may note that the ordering kinetics can be very rapid near the surface as compared to the bulk, if effective surface fields are present [4.119]. The reason is that the effective field selects only one component of the order parameter and, since this field acts on the surface plane, the situation is comparable to ordering in a one-dimensional case.

Interestingly, for the Cu_3Au type of order, for instance, the $L1_2$ structure, the nearest-neighbor interaction model already shows a rich variety of possible ordered near-surface states above the bulk transition temperature [4.118]. In the absence of surface fields as well as for strong fields of both signs, order parallel to the surface can be stabilized in the first layer ($h_1 = 0$) or the second layer (for strong h_1 favoring the majority), or buried in the third layer (if h_1 forces the minority to fill the top layer). In all these cases there are ordinary transitions in the surface layers, since unlike the CuAu type of ordering, here the surface field couples only to a subspace of the $L1_2$ order parameter. These results may be of interest for other alloys which form the $L1_2$ structure. They do not occur in Cu_3Au and none of these have been observed yet. The reason probably is that these surface ordering phenomena are more likely to occur in alloys having very-short-range interactions and that they are rather demanding to verify experimentally.

5. Summary and Conclusions

The applications presented in the previous chapter have shown that diffuse scattering experiments provide a most accurate tool to explore the pair correlation functions in alloy solid solutions. The two-point correlation functions describing the chemical short-range order and the atomic size effects are obtained by Fourier analysis methods in a straightforward manner. The use of neutrons has been of particular value, since information about the equilibrium short-range order in alloys has been obtained in in situ high-temperature experiments, while the background due to the diffuse inelastic scattering by phonons has been separated by experimental means.

The measured short-range order has been used for two purposes: first, to simulate the corresponding real-space structures and second, to determine quantitatively the effective pair interactions of real alloys. Therefore, the so-called reverse and inverse Monte Carlo methods have been applied. The advantage of Monte Carlo methods over approximate analytic approaches is that the precision of the results is limited only by the availability of computing power. The Monte Carlo methods have been discussed in detail, including possible improvements and alternatives in the algorithms. The inverse Monte Carlo method provides a unique solution for effective pair interactions, and it seems to be promising to apply this method also to the determination of interactions in liquids and amorphous systems. Pair interactions are expected to dominate in most alloys. However, the general interaction model may need to account for further contributions, e.g. possible many-body interactions, explicit magnetic exchange interactions or local volume relaxation effects. In principle, many-body interactions may be established from the variation of the pair correlation function with temperature and concentration. Interestingly, one has to conclude that the information about all many-body correlations in a single structure would not be sufficient for a unique determination. Magnetic correlations and correlations which involve lattice displacements can be explored by scattering experiments.

In this review various alloys related to the author's own work have been considered. As shown for the solubility of Co in Cu, this macroscopic information may provide a first but rather limited estimate of the atomic interactions. We have shown that interactions can be found very accurately from electronic structure calculations. The solubilities are reproduced exactly, but this has been, of course, a favorable case. By chance, further-than-second-neighbor interactions – which have not been considered although this is easily possible – either vanish or cancel each other with respect to the solubility. There is not as much information about the Cu–Co alloy system available from diffuse scattering data as for the Ni–Cu alloys. In this case an overall solid solution of Ni–Cu is observed. Heats of solution for the Ni–Cu alloy system indicate a decomposition at lower temperatures that is hindered by the decreasing atomic mobility. Diffuse scattering experiments have shown the alloy's clustering tendency in the disordered, high-temperature phase. The information about the effective interactions may be reduced slightly because only experimental results on polycrystals are available in this case. However, consistent effective pair interactions have been calculated by means of the inverse Monte Carlo method from measurements at different temperatures for at least one composition, and this provides support for the Ising-model approach to real alloys.

The iron-rich Fe–Al alloys, a basis for some superalloys, exhibit a complex and interesting interplay of magnetic and chemical order, where the magnetism may even drive the system into a new chemically ordered state. Desirable information about the magnetic correlations can be explored in future diffuse scattering experiments using polarization analysis in neutron experiments. In view of the proposed changes to the phase diagram, future small-angle scattering experiments will be useful to verify the order of the proposed low-temperature ordering transitions. At high temperatures magnetic influences on the effective pair interactions become negligible, but this does not lead to the expected quantitative agreement with observed transition temperatures at high temperatures. One may speculate that there are further important elastic energy contributions that are related to the gradient of the interaction potential. In order to clarify this point, complementary measurements with additional scattering contrast would be required to determine any related species-dependent size effects.

Some predictions of mean-field approximations and their reliability have been mentioned, e.g. their validity for the solubilities at low concentrations, and when they apply for long-range interactions; even qualitative failures may arise in calculations of interactions. Experiments on a Cu–Zn alloy and Monte Carlo simulations revealed a very simple scaling property for the temperature dependence of short-range order that is however, different from the mean-field prediction. This shows the incorrectness of the common view that the mean-field theory of short-range order (Curie–Weiss law) becomes exact in the high-temperature limit.

In Ni–Cr alloys of low Cr content, weak Fermi surface effects can be traced in the effective pair interactions determined from the scattering experiments by means of the inverse Monte Carlo method. The results are consistent with positron annihilation experiments and theoretical calculations. Interactions up to at least the fourth neighbor are necessary to describe the wave vector of the short-range order scattering and to explain the ordered ground states. No particular Fermi surface effect has been found for alloys of higher Cr concentration. The phase stability of the only known ordered Ni_2Cr phase has not yet been completely understood, neither from these experiments nor from theoretical approaches. In view of the phase stability of the alloy, any Fermi surface effect seems to be rather unimportant, unless similar effects could arise in the long-range ordered state. The observed lattice contraction upon ordering indicates important energy contributions from lattice relaxation. Simple size-effect models are not consistent with the unusual asymmetry of the diffuse scattering pattern, nor would these be capable of describing volume changes due to ordering.

The ferrous oxide $Fe_{1-x}O$ is prominent for its unavoidably large cation disorder and deviation from stoichiometry, and it has unusual ionic transport properties. Here, in situ diffuse scattering experiments on a single crystal, with various oxygen partial pressures at high temperatures, demonstrated the value of neutron techniques and yielded much insight into the defect structure. The charged defects, cation vacancies and interstitials tend to form defect clusters as predicted theoretically. In this case the pair correlations clearly confine the possible higher-order correlations. The short-range order defines the trend to form 4:1 vacancy–interstitial arrangements that can be viewed as "spinel embryos" – seeds for the neighboring spinel phase Fe_3O_4 in the homogeneous but non-stoichiometric wüstite structure. In detail, the simulated structure is determined by strongly correlated defects and has provided an understanding of the high stability of defect cluster arrangements despite the high cation mobility, and an understanding of why an increasing number of vacancies has such little effect on the cation mobility. According to the experiments, the displacement field around cation vacancies is governed by Coulomb forces, neighboring oxygen ions are attracted and next-nearest cations are repelled. In addition to structural simulation using the reverse MC method, Kanzaki model calculations proved to be very valuable, by explaining the compensation of long-range displacement fields. Surprisingly at first sight, with increasing non-stoichiometry the displacement scattering near the Bragg peaks (Huang scattering) vanishes in the experimental observation, but it is reproduced well in the model calculation by strong local correlations among the charged defects. This observation has particular relevance for ionic systems, while in metallic alloys such effects should be weaker.

The bcc metal hydrides, such as VD_x are prototypes for compact hydrogen storage systems. The diffuse scattering investigations reveal the short-range order effects that limit the storage capability of the metal. The data have provided a first quantitative determination of the mutual hydrogen interactions. The interactions are essentially repulsive, particularly for very close hydrogen neighbors, and possible contributions from elastic interactions have been discussed.

The methods are now well developed and will be useful in further applications to more complex systems, e.g. ternaries, quasicrystals, molecular solutions and polymer blends. Additional degrees of freedom will require additional contrast variation. Isotopic substitution can be used for neutron experiments and complementarily, anomalous x-ray scattering has also become a standard tool at the synchrotron sources. Atomic displacements and their role in alloy phase stability have been mentioned more in perspective views rather than in quantitative discussions. One reason is that the size effects, which are obtained in the diffuse scattering experiments, do not provide a complete picture. In general case, species-dependent displacements, due to charge transfer for instance, have also to be taken into account. First measurements, utilizing resonant anomalous x-ray scattering for contrast variation, have shown the relevance of these displacements. However, improvements in measurement and data analysis are still important for getting clearer results (for instance for the Ni–Cr alloys). Appropriate simulations that incorporate all relaxation effects will require a much larger effort, and we have only briefly discussed some first pioneering Monte Carlo simulations towards this direction.

We have also discussed important surface effects upon ordering and disordering in alloys. Continuous surface-induced disordering is a typical phenomenon that may arise in real semi-infinite alloys undergoing first-order bulk transitions, and this has been studied in highly accurate Monte Carlo simulations of a very simple model. There is an equivalence of surface induced disordering and critical wetting, and theoretical predictions have been confirmed by these Monte Carlo simulations. Universal critical exponents have been found as expected. We have also confirmed the scaling relations based on a single independent exponent determined by the interfacial roughness. A few remaining discrepancies from the theoretical predictions (concerning the thickness of the wetting layer and of the interface) were not discussed in detail but possibly rely on the non-scalar properties of the order parameter for the CuAu-type structure and on the anisotropy of the correlations even in the disordered phase.

The new phenomenon of surface-induced ordering has been found in the simulations and it is likely that this will also be found to occur in various ordering alloys. Further preliminary simulations show that even its origin, due to removed frustration, is not related only to the peculiarities of the CuAu structure. Models can be found for an analogous behavior in the CuPt, AlTi, Cu_3Au ordering types, for example. To date there are no reports on surface-induced ordering in alloys, and we hope that this work stimulates future experiments, although these may not be easy to perform and in situ conditions may be required. There has been an observation which is also very closely related to this subject: thin films of di-block copolymers between two silicon plates exhibit competing ordering phenomena which are apparently induced by the polymer–silicon interfaces [5.1]. Monte Carlo simulations and Ising(-type) model descriptions also provide a key to the microscopic understanding of heterogeneous nucleation. We have discussed the relevance of effective surface fields, which in the case of free surfaces may be expected to be determined to a large extent by the difference of mere bulk properties, namely the heats of formation (on the lattice) of the pure materials. However, relaxational energies seem to be of more importance in the particular case of Cu_3Au and $CuAu$ alloys. They presumably cause the Au enrichments at the surface and induce an alternating, layering concentration profile due to the ordering tendency of the bulk alloy.

The next step is to proceed further into the subtleties of surface effects on alloy ordering and aim at revealing modifications of the effective interactions. More detailed investigations of possible changes of the interactions near surfaces or interfaces or in thin films, similar to the bulk studies presented here, will certainly be both desirable and very challenging.

References

Chapter 1

1.1 De Fontaine D. (1979) Sol. Stat. Phys. **34**, 73–274
1.2 Khachaturyan A.G. (1983) The Theory of Structural Transformations in Solids, Wiley, New York
1.3 Ducastelle F. (1991) Order and Phase Stability in Alloys, De Boer F.R., Pettifor, D. (Eds.) North Holland
1.4 Stocks G.M., Gonis A. (Eds.) (1989) Alloy Phase Stability, NATO ASI Series E **163**, Kluwer Academic, Dordrecht
1.5 Turchi P.E.A., Gonis A. (Eds.) (1994) Statics and Dynamics of Alloy Phase Transformations, NATO-ASI Series B **319**, Plenum Press, New York
1.6 Faulkner J.S., Jordan R.G. (Eds.) (1994) Metallic Alloys: Experimental and Theoretical Perspectives, NATO-ASI Series E **256**, Kluwer Academic, Dordrecht
1.7 Gonis A., Turchi P.E.A., Kudrnovsky J. (Eds.) (1996) Stability of Materials, NATO-ASI Series B, Plenum Press, New York
1.8 Von Laue M. (1918) Ann. Phys. **56**, 497
1.9 Warren B.E., Averbach B.L. (1953) Modern Research Techniques in Physical Metallurgy, pp. 95–130, ASM, Cleveland
1.10 Guinier A. (1963) X-Ray Diffraction, Freeman, San Francisco
1.11 Chen H., Comstock R.J., Cohen J.B. (1979) Ann. Rev. Mater Sci. **9**, 51
1.12 Schwartz L., Cohen J.B. (1977) Diffraction from Materials, Academic Press, New York
1.13 Schweika W. (1994) Scattering determination of short range order in alloys pp. 103-126. In: Statics and Dynamics of Alloy Phase Transformations, Turchi P.E.A., Gonis A. (Eds.), Plenum Press, New York
1.14 Dietrich S., Fenzl W. (1989) Phys. Rev. B **39**, 8873
1.15 Dietrich S., Haase A. (1995) Scattering of X-rays and Neutrons at Interfaces, Physics Reports **260**, 1-138
1.16 Caudron R., Sarfati M., Barrachin M., Finel A., Ducastelle F., Solal F. (1992) Journal de Physique **2**, 1145
1.17 Kostorz G. (1992) In: Industrial and Technological Applications of Neutrons, p. 85, Editrice Compositori, Bologna
1.18 Schönfeld B. (1993) Habilitationsschrift, ETH Zürich
1.19 Epperson J.E., Anderson J.P., Chen H. (1994) Metall. Mat. Trans. **25**A, 17
1.20 Welberry T.R., Butler B.D. (1994) J. Appl. Cryst. **27**, 205–231
1.21 Moss S.C. (1995) Mat. Res. Soc. Symp. Proc. **376**, 675–687

Chapter 2

2.1 Messiah A. (1962) Quantum Mechanics, North Holland

2.2 Van Hove L. (1954) Phys. Rev. **95**, 249

2.3 Marshall W., Lovesey S.W. (1971) Theory of Neutron Scattering, Clarendon Press, Oxford

2.4 Squires G.L. (1978) Introduction to the Theory of Thermal Neutron Scattering, Cambridge University Press

2.5 Lovesey S.W. (1987) Theory of Neutron Scattering from Condensed Matter, Vol. I, Clarendon Press, Oxford

2.6 Ice G.E., Sparks C.J., Shaffer L.B. (1994) Resonant Anomalous X-Ray Scattering, Theory and Applications, Materlik G., Sparks C.J., Fischer K. (Eds.), pp. 265–294, Elsevier Science

2.7 Kramers H.A. (1929) Phys. Z. **30**, 522; Kronig R. de (1926) J. Opt. Soc. Amer. **12**, 547

2.8 Walker C.B., Keating D.T. (1961) Acta Cryst. **14**, 1170

2.9 Krivoglaz M.A. (1969) Theory of X-Ray and Thermal Neutron Scattering by Real Crystals (translation) Moss S.C. (Ed.), Plenum Press, New York

2.10 Kubo R. (1962) J. Phys. Soc. Japan **17**, 1100

2.11 Faber T.E., Ziman J.M. (1965) Phil. Mag. **11**, 153

2.12 Bathia A., Thornton D.E. (1970) Phys. Rev. B **2**, 3004

2.13 Hayakawa M., Cohen J.B. (1975) Acta Cryst. A **31**, 635

2.14 Schweika W., Hoser A., Martin M., Carlsson A.E. (1995) Phys. Rev. B **51**, 15771–15788

2.15 Pionke M. (1995) Thesis, RWTH Aachen

2.16 Matsubara T.J. (1952) J. Phys. Soc. Japan **7**, 270

2.17 Kanzaki H. (1957) J. Phys. Chem. Solids **2**, 24; Kanzaki H. (1957) J. Phys. Chem. Solids **2**, 107

2.18 Cook H.E. (1969) J. Phys. Chem. Solids **30**, 1097

2.19 Trinkaus H. (1972) Phys. Stat. Sol. B **51**, 307

2.20 Dederichs P. (1973) Journal de Physique **3**, 471

2.21 Vegard L. (1921) Z. Phys. **5**, 17

2.22 Sparks C.J., Borie B. (1966) Local atomic arrangements studied by x-ray diffraction, pp. 5–46, Gordon & Breach, New York

2.23 Borie B., Sparks C.J. (1971) Acta Cryst. A**27**, 198

2.24 Williams R.O. (1974) Metall. Trans. **5**, 1843

2.25 Lawson C.L., Hanson R.J. (1974) Solving Linear Least Squares Problem, Prentice Hall, Englewood Cliff, New Jersey

2.26 Tibballs J.E. (1975) J. Appl. Cryst. **8**, 11

2.27 Georgopoulos P., Cohen J.B. (1977) Journal de Physique C **38**, 191; Georgopoulos P., Cohen J.B. (1981) Acta Metall. **29**, 1535; Auvray X., Georgopoulos P., Cohen J.B. (1981) Acta Metall. **29**, 1061

2.28 Ice G.E., Sparks C.J., Habenschuss A., Shaffer L.B. (1992) Phys. Rev. Lett. **68**, 863

2.29 Jiang X., Ice G.E., Sparks C.J., Robertson L. , Zschak P. (1996) Phys. Rev. B **54**, 3211

2.30 Goff J.P., Hutchings M.T., Brown K., Hayes W., Godfrin H. (1990) Mat. Res. Soc. Symp. Proc. **166**, 373–377

2.31 Schuster B., Dieckmann R., Schweika W. (1989) Ber. Bunsenges. Phys. Chem. **93**, 1347

2.32 Hashimoto S. (1974) Acta Cryst. A **30**, 792

2.33 Neder R.B., Frey F., Schulz H. (1990) Acta Cryst. A **46**, 792

2.34 Huang K. (1947) Proc. Roy. Soc. A **190**, 102

2.35 Ehrhart P. (1994) J. Nucl. Mater. **216**, 170–198

2.36 Bauer G.S., Seitz E., Just W. (1975) J. Appl. Cryst. **8**, 162

2.37 Schmatz W. (1973) X-ray and neutron scattering studies in disordered crystals. In: Treatise in Materials Science and Technology, Vol.2, pp. 105–229, Herman H.H. (Ed.), Academic Press, New York; Schmatz W. (1978) Diffuse scattering. In: Neutron Diffraction, Topics in Current Physics 6, H. Dachs (Ed.), Springer, Berlin

2.38 Ehrhart P., Haubold H.G., Schilling W. (1974) Adv. Solid State Phys. **14**, 87

2.39 Gehlen P.C., Cohen J.B. (1965) Phys. Rev. **139**, 844

2.40 Schärpf O., Gabrys B., Pfeiffer D.G. (1990) ILL–Report 90SC26T, Institut Laue–Langevin, Grenoble

2.41 Lamers C., Schärpf O., Schweika W., Batoulis J., Sommer K., Richter D. (1992) Physica B **180** & **181**, 495

2.42 Internet address: http://www.kfa-juelich.de/iff/iff_sm.e.html (→ DNS)

2.43 Reinhard L., Robertson J.L., Moss S.C., Ice G.E., Zschack P., Sparks C.J. (1992) Phys. Rev. B **45**, 2662

2.44 Dosch H. (1992) Critical Phenomena at Surfaces and Interfaces (Evanescent X-ray and Neutron Scattering), Springer Tracts in Modern Physics Vol. 126, Höhler G. (Ed.), Springer, Berlin

2.45 Schlegel P., Hadfield R.A., Casalta H., Andersen N.H., Poulsen H.F., von Zimmermann M., Schneider J.R., Ruixing Liang, Dosanjih P., Hardy W.N. (1995) Phys. Rev. Lett. **74**, 1446

Chapter 3

3.1 Ising E. (1925) Beitrag zur Theorie des Ferromagnetismus, Z. F. Phys. **31**, 253

3.2 Kikuchi R. (1951) Phys. Rev. **81**, 988

3.3 Sanchez J.M., Ducastelle F., Gratias D. (1984) Physica A **128**, 334

3.4 Mohri T., Sanchez J.M., de Fontaine D. (1985) Acta Metall. **33**, 1171–85

3.5 Finel A. (1994) in Statics and Dynamics of Alloy Phase Transformations, (Eds.) Turchi P.E.A., Gonis A., NATO ASI Series B, Physics Vol. 319, Plenum Press, New York p.495

3.6 Binder K. (Ed.) (1979) Monte Carlo Methods in Statistical Physics, Springer, Berlin; Binder K. (Ed.) (1984) Applications of the Monte Carlo Method in Statistical Physics, Springer, Berlin; Binder K. (Ed.) (1986) The Monte Carlo Method in Statistical Physics, 2nd edition, Springer, Berlin; Binder K. (Ed.) (1987) Applications of the Monte Carlo Method in Statistical Physics, 2nd edition, Springer, Berlin; Binder K. (Ed.) (1991) The Monte Carlo Method in Condensed Matter Physics, Springer, Berlin; Binder K., Heermann D.W. (Eds.) (1992) The Monte Carlo Simulation in Statistical Physics, 2nd edition, Springer, Berlin; Binder K. (Ed.) (1995) The Monte Carlo Method in Condensed Matter Physics, 2nd edition, Springer, Berlin

3.7 Mouritsen O.G. (1984) Computer Simulations of Phase Transitions and Critical Phenomena, Springer, Heidelberg

3.8 Binder K. (1985) J. Comput. Phys. **59**, 1–55

3.9 Heermann D.W. (1986) Introduction to the Computer Simulation Methods of Theoretical Physics, Springer, Heidelberg

3.10 Binder K. (1986) In: Festkörperprobleme (Advances in Solid State Physics), Grosse P. (Ed.), Vol. 26, pp. 133–168, Vieweg, Braunschweig

114   References

3.11 Binder K. (1987) In: International Meeting on Advances on Phase Transitions and Disorder Phenomena, Amalfi, June 1986, Busiello G., De Cesare L., Manchini F., Marinaro M. (Eds.), pp. 1-71, World Scientific, Singapore
3.12 Binder K., Heermann D.W. (1988) The Monte Carlo Method in Statistical Physics. An Introduction, Springer, Berlin
3.13 Privman V. (Ed.) (1990) Finite Size Scaling and the Numerical Simulation of Statistical Systems, World Scientific, Singapore
3.14 Binder K. (1992) In: Computational Methods in Field Theory, Lang C.B., Gausterer A. (Eds.), p. 59, Springer, Berlin
3.15 Schweika W. (1992) Monte Carlo simulations of order–disorder phenomena in binary alloys. In: Structural and Phase Stability of Alloys, Morán-López J.L., Mejía-Lira F., Sanchez J.M. (Eds.), Plenum Press, New York
3.16 Chakraborty B. (1995) Europhys. Lett. **30**, 531
3.17 Hume-Rothery W. (1983) In: Electrons, Atoms, Metals, and Alloys, 3rd ed., Part 1, Pergamon, New York
3.18 Car R., Parinello M. (1985) Phys. Rev. Lett. **55**, 2471
3.19 Payne M.C., Teter M.P., Allan D.C., Arias T.A., Joannopoulos J.D. (1992) Rev. Mod. Phys. **64**, 1045
3.20 Chelikowsky J.R., Jing X., Wu K., Saad Y. (1996) Phys. Rev. B **53**, 12071
3.21 Stoltze P. (1997) Simulation Methods in Atomic-Scale Materials Physics, Polyteknisk Forlag, DK-2800 Lyngby, Denmark
3.22 Mousseau N., Thorpe M.F. (1992) Phys. Rev. B 45, 2015
3.23 Braspenning P.J., Zeller R., Lodder A., Dederichs P.H. (1984) Phys. Rev. B **29**, 703
3.24 Drittler B., Weinart M., Zeller R., Dederichs P.H. (1989) Phys. Rev. B **39**, 930
3.25 Connolly J.W.D., Williams A.R. (1983) Phys. Rev. B **27**, 5169–5172
3.26 Zunger A. (1994) In: Statics and Dynamics of Alloy Phase Transformations, Turchi P.E.A., Gonis A. (Eds.), p. 361, NATO ASI Series B, Physics Vol. 319, Plenum Press, New York
3.27 Pauling L. (1938) Phys. Rev. **54**, 899
3.28 Heine V., Hafner J. (1990) In: Many-Atom Interactions in Solids Nieminen R.M., Puska M.J., Manninen J.M. (Eds.), p. 12–34, Springer, Berlin
3.29 Heine V. (1980) Solid State Phys. **35**, 1
3.30 Brout R. (1965) Phase Transitions, Benjamin, New York
3.31 Clapp P.C., Moss S.C. (1966) Phys. Rev. **142**, 418; (1968) Phys. Rev. **171**, 754–761; Moss S.C., Clapp P.C. (1968) Phys. Rev. **171**, 762–777
3.32 Tokar V.I. (1985) Phys. Lett. **110A**, 453
3.33 Tokar V.I., Masanskii I.V., Grishchenko (1990) J. Phys. Condens. Matter **2**, 10199
3.34 Masanskii I.V., Tokar V.I., Grishchenko T.A. (1991) Phys. Rev. B **44**, 4647
3.35 Reinhard L., Moss S.C. (1993) Ultramicroscopy **52**, 223
3.36 Kanamori J., Kakehashi Y. (1977) Journal de Physique C **38**, 274
3.37 Allen S.M., Cahn J.W. (1973) Scripta Metall. **7**, 1261
3.38 Metropolis N., Rosenbluth A.W., Rosenbluth M.N., Teller A.H., Teller E. (1953) J. Chem. Phys. **21**, 108
3.39 Glauber R.J. (1963) J. Math. Phys. **4**, 294
3.40 Kawasaki K. (1972) In: Phase Transitions and Critical Phenomena, Domb C., Green M.S. (Eds.), Vol. 2, p. 443, Academic Press, New York
3.41 Creutz M. (1987) Phys. Rev. D **36**, 515
3.42 Adler S.L. (1988) Phys. Rev. D **38**, 1349
3.43 Klemradt U., Drittler B., Weinart M., Zeller R., Dederichs P.H. (1990) Phys. Rev. Lett. **64**, 2803

3.44 Klemradt U., Drittler B., Hoshino T., Zeller R., Dederichs P.H., Stefanou N. (1991) Phys. Rev. B **43**, 9487

3.45 Swendsen R.H., Wang J.-S. (1987) Phys. Rev. Lett. **58**, 86

3.46 Fisher M.E. (1971) Proceedings of the International Summer School *Enrico Fermi*, Course 51, Academic Press, New York

3.47 Fisher M.E. (1971) In: Critical Phenomena, p. 5, Green M.S. (Ed.) Academic Press, New York

3.48 Barber M.N. (1983) In: Phase Transitions and Critical Phenomena, Domb C., Lebowitz J.L. (Eds.), Vol.8, p.145, Academic Press, New York

3.49 Privman V., Fisher M.E. (1983) J. Stat. Phys. **33**; (1985) Phys. Rev. B **32**, 447

3.50 Binder K. (1981) Z. Phys. B **43**, 119

3.51 Binder K., Landau D.P. (1984) Phys. Rev. B **30**, 1477–1485

3.52 Binder K. (1987) Rep. Prog. Phys. **50**, 783

3.53 Binder K. (1981) Z. Phys. B **45**, 61

3.54 Lim S.H., Hasebe M., Murch G.E., Oates W.A. (1990) Phil. Mag. B **62**, 173

3.55 Hüller A. (1994) Z. Physik B **93**, 401

3.56 Wansleben S. (1987) Comp. Phys. Comm. **43**, 315

3.57 Ito N., Kanada Y. (1988) Supercomputer 25, Vol. V-**3**, 31–48

3.58 Heuer H.O. (1990) Europhys. Lett. **12**, 551

3.59 Kirkpatrick S., Stoll E. (1981) J. Comp. Phys. **40**, 517

3.60 James F. (1980) Repts. Prog. Phys. **43**, 1145

3.61 Selke W., Tapalov A.L., Shchur L.N. (1993) JETP Letters **58**, 665–668

3.62 Makino J., Miyamura O. (1995) Parallel Computing **21**, 1015–1028

3.63 Coddington P.D. (1994) Int. J. Mod. Phys. C [Physics and Computers] **5**, 547–560

3.64 Ferrenberg A.M., Landau D.P., Wong Y.J. (1992) Phys. Rev. Lett. **69**, 3382–3384

3.65 Clapp P.C. (1971) Phys. Rev. B **4**, 255

3.66 Renniger A.L., Rechtin M.D., Averbach B.L. (1974) J. Non-Cryst. Solids **16**, 1

3.67 Faulkner J.S. (1985) private communication

3.68 McGreevy R.L., Pusztai L. (1988) Mol. Sim. **1**, 359

3.69 Gerold V., Kern J. (1987) Acta Metall. **35**, 393

3.70 Livet F. (1987) Acta Metall. **35**, 2915

3.71 Schweika W. (1989) Effective pair interactions in binary alloys. In: Alloy Phase Stability, Stocks G.M., Gonis A. (Eds.), NATO ASI Series E 163, Kluwer Academic, Dordrecht

3.72 Kern J. (1983) Thesis, Univ. Stuttgart

3.73 Schweika W., Haubold H.G. (1988) Phys. Rev. B **37**, 9240

3.74 Schweika W. (1985) Thesis, RWTH Aachen

3.75 Schweika W. (1990) Mat. Res. Soc. Symp. Proc. **166**, 249

3.76 Schönfeld B., Reinhard L., Kostorz G. (1988) Phys. Stat. Sol. B **147**, 457

3.77 Schönfeld B., Ice G.E., Sparks C.J., Haubold H.G., Schweika W., Shaffer L.B. (1994) Phys. Stat. Sol. B **183**, 79

3.78 Ostheimer M., Bertagnolli H. (1989) Mol Sim. **3**, 227

3.79 Bieber A., Gautier F. (1984) J. Phys. Soc. Japan **53**, 2061

3.80 Pionke M., Schweika W., Springer T., Sonntag R., Hohlwein D. (1995) Physica Scripta **T57**, 107–111

3.81 Welberry T.R., Withers R.L. (1991) J. Appl. Cryst. **24**, 18–29

3.82 Evans R.A. (1990) Mol. Sim. **4**, 409

3.83 Henderson R.L., (1974) Phys. Lett. **49A**, 197

3.84 Casanova G., Dulla R.J., Jonah D.A., Rowlinson J.S., Saville G. (1970) Mol. Phys. **18**, 589–606

3.85 Rowlinson J.S. (1984) Mol. Phys. **52**, 567

3.86 Taggart G.B., Tahir-Keli R.A. (1972) Prog. Theor. Phys. **47**, 370

3.87 Schweika W., Carlsson A.E. (1989) Phys. Rev. B **40**, 4990

3.88 Wolverton C., Zunger A., Schönfeld B. (1997) Solid Sate Commun. **101**, 519

3.89 Kikuchi R., de Fontaine D. (1978) In: Applications of Phase Diagrams in Metallurgy and Ceramics, National Bureau of Standards, Special Publication No. 496, Vol. 2, p. 967, US Government Printing Office, Washington, DC; de Fontaine D., Kikuchi R., ibid., p. 999

3.90 Ornstein L.S., Zernike F. (1914) Proc. Acad. Sci. Amsterdam **17**, 793; (1918) Z. Physik **19**, 134; (1926) Z. Physik **27**, 761

3.91 Fisher M.E. , Burford R.J. (1967) Phys. Rev. **156**, 583

3.92 Barocchi F., Chieux P., Reatto L., Tau M. (1993) Phys. Rev. Lett. **70**, 947 (1988) Phys. Rev. B **37** 16, 9240

Chapter 4

4.1 Ferrenberg A.M., Landau D.P. (1991) Phys. Rev. B **44**, 5081

4.2 Finel A., private communications

4.3 Sanchez J.M., de Fontaine D. (1978) Phys. Rev. B **17**, 2926

4.4 Burley D.M. (1972) In: Phase Transitions and Critical Phenomena, Domb C., Green M.S. (Eds.) Vol. 2, p. 329, Academic Press, New York

4.5 Binder K. (1989) Mechanisms for the decay of unstable and metastable phases: spinodal decomposition, nucleation and late-state coarsening, in Alloy Phase Stability Stocks G.M., Gonis A. (Eds.), p. 233–268, NATO ASI Series E, Kluwer, Dordrecht

4.6 Tammann G., Oelsen W. (1930) Z. Anorg. Allg. Chem. **186**, 257

4.7 Nishizawa T., Ishida K. (1984) Bull. Alloy Phase Diagrams **5** (2), 161–165

4.8 Hoshino T., Schweika W., Zeller R., Dederichs P.H. (1993) Phys. Rev. B **47**, 5106–5117

4.9 Mozer B., Keating D.T., Moss S.C. (1968) Phys. Rev. B **175**, 868

4.10 Vrijen J., van Royen E.W., Hoffman D.W., Radelaar S. (1977) Journal de Physique C**7**, 187; J. Vrijen, S. Radelaar (1978) Phys. Rev. B **17**, 409

4.11 Wagner W., Poerschke R., Axmann A., Schwahn D. (1980) Phys. Rev. B **21**, 3087

4.12 Hansen M., Anderko K. (Eds.) (1958) Constitution of Binary Alloys, McGraw-Hill, New York; Elliot R.P. (1965) Constitution of Binary Alloys: First Supplement, McGraw-Hill, New York; Shunk F.A. (1969) Constitution of Binary Alloys: Second Supplement, McGraw-Hill, New York; (1980) Handbook of Binary Phase Diagrams, Genium, Schenectady, New York; Massalski T.B. (1990) Binary Alloy Phase Diagrams, 2nd ed., ASM International, New York

4.13 Chen L.Q., Wang Y.Z., Khachaturyan A.G. (1994) In: Statics and Dynamics of Alloy Phase Transformations, Turchi P.E.A., Gonis A. (Eds.), p.587 Plenum Press, New York

4.14 Fratzl P., Penrose O. (1995) Acta Metall. et Mater. **43**, 2921–2930

4.15 Semenovskaya S.V. (1974) Phys. Stat. Sol. B **64**, 291

4.16 Pierron-Bohnes V., Cadeville M.C., Finel A., Schärpf O. (1991) J. Phys. Condens. Matter. **1**, 247

4.17 Contreras-Solorio D.A., Mejía-Lira F., Morán-López J.L., Sanchez J.M., (1988) Journal de Physique C8, 105; Contreras-Solorio D.A., Mejía-Lira F., Morán-López J.L., Sanchez J.M. (1988) Phys. Rev. B **38**, 11481

4.18 Dünweg B., Binder K. (1987) Phys. Rev. B **36**, 6935

4.19 Bichara C., Inden G. (1991) Scripta Metall. **25**, 2607

4.20 Schmid F., Binder K. (1992) J. Phys. Condens. Matter **4**, 3569

4.21 Cable J.W., Werner S.A., Felcher G.P., Wakabayashi N. (1984) Phys. Rev. B **29**, 1268–1278

4.22 Lawley A., Cahn R.W. (1961) J. Phys. Chem. Solids **20**, 204–221

4.23 Fultz B., Gao Z.Q. (1993) Phil. Mag. B **67**, 787

4.24 Becker M. (1995) Diplom, RWTH Aachen; Becker M., Schweika W. (1996) Scripta Metall. **35**, 1259

4.25 Allen S.M., Cahn J.W. (1976) Acta Metall. **24**, 425

4.26 Oki K., Hasada M., Eguchi T. (1973) Jap. J. Appl. Phys. **10**, 1522

4.27 Köster W., Gödecke T. (1980) Z. Metallkde. **71**, 765

4.28 Kubashewski O. (Ed.) (1982) Iron-Binary Phase Diagrams, Springer, Berlin

4.29 Inden G., Pepperhoff W. (1990) Z. Metallkde. **81**, 770

4.30 Vintaykin Ye.Z., Loshmanov A.A. (1967) Fiz. met. metalloved **27**, (7), 754; Vintaykin Ye.Z., Urushadze G.G. (1969) Fiz. met. metalloved **27**, (5), 895

4.31 Cowley J.M. (1950) Phys. Rev. **77**, 669

4.32 Turchi P.E.A., Pinski F.J., Howell R.H., Wachs A.L., Fluss M.J., Johnson D.D., Stocks G.M., Nicholson D.M., Schweika W. (1990) Mat. Res. Soc. Symp. Proc. **166**, 231

4.33 Staunton J.B., Johnson D.D., Pinski F.J. (1994) Phys. Rev. B **50**, (3), 1450–1472

4.34 Ice G.E., Sparks, Jr. C.J. (1990) Nucl. Inst. Meth. Phys. Res. A **291**, 110

4.35 Stanley H.E. (1971) In: Critical Phenomena in Alloys, Magnets and Superconductors, Mills R.E., Ascher E., Jaffee R.I. (Eds.), McGraw-Hill, New York

4.36 Ginzburg V.L. (1960) Fiz. Tverd. Tela. **2**, 2031; (1960) Sov. Phys. Solid State **2**, 1824

4.37 Arrott A.S. (1985) Phys. Rev. B **31**, 2851

4.38 Fähnle M., Souletie J. (1984) Journal de Physique C17, L469; (1985) Phys. Rev. B **32**, 3328; (1986) Phys. Stat. Sol. **138**, 181

4.39 Dietrich O.W., Als-Nielsen J. (1967) Phys. Rev. **153**, 711

4.40 Lamers C., Schweika W. (1992) Physica B 180&181 326

4.41 Turchi P.E.A., Sluiter M., Pinski F.J., Nicholson D.M., Stocks G.M., Staunton J.B. (1991) Phys. Rev. Lett. **67**, 1779

4.42 Belyakov M.Y., Kiselev S.B. (1992) Physica A **190**, 75

4.43 Anisimov M.A., Kiselev S.B., Sengers V.J., Tang S. (1992) Physica A **188**, 487

4.44 Meier G., Schwahn D., Mortensen K., Janssen S. (1993) Europhys. Lett. **22**, 577–583

4.45 Sengers J.V. (1994) Effects of critical fluctuations on the thermodynamic and transport properties of supercritical fluids, In: Supercritical Fluids, pp. 231–271, Kluwer Academic, Netherlands

4.46 Schweika W., Hoser A., Martin M., Carlsson A.E. (1995) Phys. Rev. B **51**, 15771–15788

4.47 Muan A., Osborn E.F. (1965) Phase Equilibria among Oxides in Steelmaking, Addison-Wesley, Reading, Massachusetts; see also Darken L.S., Gurry R.W. (1945) J. Am. Ceram. Soc **67**, 1398

4.48 Rekas M., Mrowec S. (1987) Solid State Ionics **22**, 185

4.49 Cheetham A.K., Fender B.E.F., Taylor R.I. (1971) J. Phys. C: Solid State Phys. **4**, 2160

4.50 Radler M., Cohen J.B., Faber Jr. J. (1990) J. Phys. Chem. Solids **51**, 217
4.51 Catlow C.R.A., Stoneham A.M. (1981) J. Am. Ceram. Soc **64**, 234
4.52 Grimes R.W., Anderson A.B., Heuer A.H. (1986) J. Am. Ceram. Soc **69**, 619
4.53 Press M.R., Ellis D.E. (1987) Phys. Rev. B **35**, 9, 4438
4.54 Stoneham A.M. (1980) Physics Today **33**, 34
4.55 Chen W.K., Peterson N. (1975) J. Phys. Chem. Solids **36**, 1097
4.56 Koch F., Cohen J.B. (1969) Acta Cryst. B **25**, 275
4.57 Hayakawa M., Morinaga M., Cohen J.B. (1973) In: Defects and Transport in Oxides, Seltzer M.S., Faffee R.I., (Eds.), p.177, Plenum Press, New York
4.58 Gartstein E., Mason T.O., Cohen J.B. (1986) J. Phys. Chem. Solids **47**, 759
4.59 Kugel G., Carabatos C., Hennion B., Prevot B., Revcolevschi A., Tochetti D. (1977) Phys. Rev. B **16**, 378
4.60 Hoshen J., Kopelman R. (1976) Phys. Rev. B **14**, 3428
4.61 Roth W.L. (1960) Acta Cryst. **13**, 140
4.62 Monty C. (1989) In: Diffusion in Materials Lasker A.L., Bocquet J.L., Monty C., NATO-ASI Series E: Applied Science, Vol.179, Kluwer Academic, Dordrecht
4.63 Hoser A., Martin M., Schweika W., Carlsson A.E., Caudron R., Pyka N. (1994) Solid State Ionics **72**, 72–75
4.64 Hoser A., Martin M., Schweika W., Carlsson A.E., Caudron R., Glas R., in preparation.
4.65 Radler M.J., Cohen J.B., Sykora J.P., Mason T., Ellis D.E., Faber J.F. (1992) J. Phys. Chem. Solids **53**, 141
4.66 Tetot R., Nacer B., Boureau G. (1993) J. Phys. Chem. Solids **54**, 517–525
4.67 Lacher J.R. (1937) Proc. Roy. Soc. A **161**, 525
4.68 Alefeld G. (1969) Phys. Stat. Sol. **32**, 67
4.69 Alefeld G. (1972) Ber. Bunsenges. Phys. Chem. **76**, 746
4.70 Oates W.A., Hasebe M., Meuffels P., Wenzl H. (1985) Z. Phys. Chem. Neue Folge **146**, 201
4.71 Alefeld G., Voelkl 1978 In: Topics in Applied Physics: Hydrogen in Metals Vols. 28 and 29, Springer, Berlin
4.72 Fukai Y. (1993) The Metal Hydrogen System, Springer Series in Materials Science, Vol. 21, Springer, Berlin, Heidelberg, New York
4.73 Hempelmann R., Richter D., Faux D.A., Ross D.K. (1988) Z. Phys. Chem. Neue Folge **159**, 175
4.74 Knell U., Wipf H., Lautenschläger G., Hock R., Weitzel H., Ressouche E. (1994) J. Phys. Condens. Matter **6**, 1461
4.75 Horner H., Wagner H. (1974) J. Phys. C **7**, 3305
4.76 Pionke M., Schweika W., Springer T., Sonntag R., D. Hohlwein (1995) Physica B **213&214**, 567–569; elastic interactions are erroneously displayed too small by a factor 10
4.77 Cheryakov A.Y., Entin I.R., Somenkov V.A., Shil'shtein S.Sh., Chertkov A.A. (1972) Sov. Phys. Solid State **13**, 2172; Somenkov V.A., Entin I.R., Cheryakov A.Y., Shil'shtein S.Sh., Chertkov A.A. (1972) Sov. Phys. Solid State **13**, 2178
4.78 Siems R. (1968) JÜL-Report des Forschungszentrums Jülich, Nr. 545-FN
4.79 Blanter M.S., Khachaturyan A.G. (1978) Metall. Trans. A **9a**, 753
4.80 Mokrani A., Demangeat C., Moriatis G. (1987) J. Less-Common Met. **130**, 209
4.81 Leibfried G., Breuer N. (1978) Point Defects in Metals I, Springer Tracts in Modern Physics Vol. 81, Springer, Heidelberg
4.82 Sak J. (1973) Phys. Rev. B **8**, 281
4.83 Fisher M.E., Ma S.-K., Nickel B.G. (1972) Phys. Rev. Lett. **29**, 917

4.84 Folk R., Moser G. (1994) Phys. Rev. E **49**, 3128

4.85 Schwahn D., Schmatz W. (1978) Acta Metall. **26**, 1653

4.86 De Gironcoli S., Giannozzi P., Baroni S. (1991) Phys. Rev. Lett. **66**, 2116; Baroni S., de Gironcoli S., Giannozzi P. (1992) Structural and Phase Stability of Alloys, Morán-López J.L., Mejía-Lira F., Sanchez J.M. (Eds.), Plenum Press, New York

4.87 Wolverton C., Zunger A. (1996) In: Stability of Materials, NATO-ASI Series B **355**, Gonis A., Turchi P.E.A., Kudrnovsky J. (Eds.), Plenum Press, New York

4.88 Dünweg B., Landau D.P. (1992) Phys. Rev. B **48**, 14182

4.89 Laradji M., Landau D.P., Dünweg B. (1995) Phys. Rev. B **51**, 4894

4.90 Vandeworp E.M., Newman K.E. (1995) Phys. Rev. B **52**, 4086

4.91 Schweika W., Landau D.P., Binder K. (1996) Phys. Rev. B **53**, 8937

4.92 Schweika W., Landau D.P., Binder K. (1996) Stability of Materials, NATO-ASI Series B **355**, Gonis A., Turchi P.E.A., Kudrnovsky J. (Eds.), Plenum Press, New York , p. 165

4.93 Landau D.P., Schweika W., Binder K. (1996) In: Thin Films and Phase Transitions on Surfaces II, Michailov M. (Ed.), Coral Press

4.94 Dietrich S. (1988), In: Phase Transitions and Critical Phenomena, Domb C., Lebowitz J.L. (Eds.), Vol. 12, Chap. 1, Academic Press, New York

4.95 Pluis B., Denier van der Gon A.W., van der Veen J.F., A.J. Riemersma (1990) Surf. Sci. **239**, 265; Pluis B., Frenkel P., van der Veen J.F. (1990) Surf. Sci. **239**, 282

4.96 Lipowsky R. (1987) Ferroelectrics **73**, 69; see also Lipowsky R. (1984) J. Appl. Phys. **55**, 2485

4.97 Lipowsky R., Kroll D.M., Zia R.K. (1983) Phys. Rev. B **27**, 4499

4.98 Binder K., Landau D.P. (1992) Phys. Rev. B **46**, 4844

4.99 Binder K., (1996) In: Stability of Materials, NATO-ASI Series B **355**, Gonis A., Turchi P.E.A., Kudrnovsky J. (Eds.), Plenum Press, New York

4.100 Sundaram V.S., Farrell B., Alben R.S., Robertson W.D. (1973) Phys. Rev. Lett. **31**, 1136; McRae E.G., Malic R.A. (1984) Surf. Sci. **148**, 551; Alvarado S.F., Campagna M., Fattah A., Uelhoff W. (1987) Z. Phys. B **66**, 103

4.101 Zhu X.-M., Zabel H., Robinson I.K., Vlieg E., Dura J.A., Flynn P.C. (1990) Phys. Rev. Lett. **65**, 2692

4.102 Burandt B., Press W., Haussühl S. (1993) Phys. Rev. Lett. **71**, 1188

4.103 Ricolleau C., Loiseau A., Ducastelle F., Caudron R. (1992) Phys. Rev. Lett. **68**, 3591; Loiseau A., Ricolleau C., Cabet E. (1993) Journal de Physique IV, C**7**, (3) 2057–2062; see also Morris D.G. (1975) Phys. Stat. Sol. (a) **32**, 145; Leroux C., Loiseau A., Cadeville M.C., Broddin D., Tendeloo G. (1990) J. Condens. Matter **2**, 3479

4.104 Kikuchi R., Cahn J.W. (1979) Acta Metall. **27**, 1337

4.105 Cahn J.W., Hilliard J.C. (1958) J. Chem. Phys. **28**, 258

4.106 Boulter C.J., Parry A.O. Phys. Rev. Lett. (1995), 3403

4.107 Schweika W., Binder K., Landau D.P. (1990) Phys. Rev. Lett. **65**, 3321

4.108 Binder K., (1983) In: Phase Transitions and Critical Phenomena Domb C., Lebowitz J.L. (Eds.), Vol. 8 Chap. 1, Academic Press, New York

4.109 Schwartz L.H., Cohen J.B. (1965) J. Appl. Phys. **36**, 598

4.110 Kroll D.M., Gompper G. (1987) Phys. Rev. B **36**, 7078

4.111 Gompper G., Kroll D.M. (1988) Phys. Rev. B **38**, 459

4.112 F. Schmid (1996) In: Stability of Materials, NATO-ASI Series B **355**, Gonis A., Turchi P.E.A., Kudrnovsky J. (Eds.), Plenum Press, New York

4.113 Reichert H., Eng P.J., Dosch H., Robinson I.K. (1995) Phys. Rev. Lett. **70**, 2006; see also Robinson I.K., Eng P.J. (1995) Phys. Rev. B **52**, 9955–9963

4.114 Mecke K.R., Dietrich S. (1995) Phys. Rev. B **52**, 2107–2116

4.115 Brewer L., Rosenblatt G.M. (1969) High Temp. Chem. **2**, 1

4.116 Livet F., Bessiere M. (1987) J. Physique **48** 1703–1708

4.117 Butler B.D., Cohen J.B. (1989) J. Appl. Phys. **65**, 2214

4.118 Schweika W., Landau D.P. (1997) In: Computer Simulation Studies in Condensed Matter Physics X, D.P. Landau, K.K. Mon, H.-B. Schuettler (Eds.), Springer, Heidelberg

4.119 Reichert H., Eng P.J., Dosch H., Robinson I.K. (1997) Phys. Rev. Lett. **78** 3475

Chapter 5

5.1 Russel T.P., Lambooy P., Kellogg G.J., Mayes A.M. (1995) Physica B **213&214**, 22–25

Index

Springer Tracts in Modern Physics